Mitsubishi FX Programmable Logic Controllers

Mitsubishi FX Programmable Logic Controllers

Applications and Programming

Second edition

JOHN RIDLEY
Diploma in Electrical Engineering, C.Eng., MIEE.

PLC Consultant MFI Manufacturing Runcorn
Cheshire

AMSTERDAM · BOSTON · HEIDELBERG · LONDON
NEW YORK · OXFORD · PARIS · SAN DIEGO
SAN FRANCISCO · SINGAPORE · SYDNEY · TOKYO

Newnes is an imprint of Elsevier

Newnes
An imprint of Elsevier
Linacre House, Jordan Hill, Oxford OX2 8DP
200 Wheeler Road, Burlington, MA 01803

First published 1997
Second edition 2004

Copyright © 1997, 2004, John Ridley. All rights reserved

The right of John Ridley to be identified as the author of this work has been
asserted in accordance with the Copyright, Designs and Patents Act 1988

No part of this publication may be reproduced in any material form (including
photocopying or storing in any medium by electronic means and whether
or not transiently or incidentally to some other use of this publication) without
the written permission of the copyright holder except in accordance with the
provisions of the Copyright, Designs and Patents Act 1988 or under the terms of
a licence issued by the Copyright Licensing Agency Ltd, 90 Tottenham Court Road,
London, England W1T 4LP. Applications for the copyright holder's written
permission to reproduce any part of this publication should be addressed
to the publisher

Permissions may be sought directly from Elsevier's Science & Technology Rights
Department in Oxford, UK: phone: (+44) 1865 843830, fax: (+44) 1865 853333,
e-mail: permissions@elsevier.co.uk. You may also complete your request on-line via
the Elsevier homepage (http://www.elsevier.com), by selecting 'Customer Support'
and then 'Obtaining Permissions'

British Library Cataloguing in Publication Data
A catalogue record for this book is available from the British Library

Library of Congress Cataloguing in Publication Data
A catalogue record for this book is available from the Library of Congress

ISBN 0 7506 56794

For information on all Newnes publications
visit our website at http://books.elsevier.com

Typeset by Integra Software Services Pvt. Ltd, Pondicherry, India
www.integra-india.com
Printed and bound in Great Britain

Working together to grow
libraries in developing countries

www.elsevier.com | www.bookaid.org | www.sabre.org

ELSEVIER   BOOK AID International   Sabre Foundation

To my wife Greta
Without her continual support, I would never have completed this work.

In Memory
This book is dedicated to the memory of Danny Bohane
of Honda of the UK Manufacturing Ltd. Swindon,
who died aged 42, June 2001.

His teaching of PLC fault-finding techniques,
I and many others will never forget.

Contents

Preface — xv
Acknowledgements — xvii
Resources — xix
Glossary — xxi

1 Introduction to PLCs — 1
   1.1 Basic PLC units — 1
   1.2 Comparison of PLC and RELAY systems — 2
   1.3 PLC software — 2
   1.4 Gx-Developer software — 3
   1.5 Hardware configuration — 3
   1.6 Base unit, extension units and extension blocks — 4
   1.7 PLC voltage supplies — 4
   1.8 Smaller FX2N PLCs — 4
   1.9 Larger FX2N PLCs — 6
   1.10 5 V DC supply — 6
   1.11 Special unit power supply requirements — 6
   1.12 Part number — 7
   1.13 Serial Number — 7
   1.14 PLC inputs — 8
   1.15 AC inputs — 8
   1.16 PLC outputs — 9
   1.17 Source–sink inputs — 10
   1.18 The source/sink – S/S connection — 11
   1.19 Source inputs – block diagram — 11
   1.20 Sink inputs – block diagram — 12
   1.21 Proximity sensors — 12
   1.22 S/S terminal configurations — 13
   1.23 PLC ladder diagram symbols — 13
   1.24 PLC address ranges — 15
   1.25 Basic operation of a PLC system — 15
   1.26 Block diagram – basic operation of a PLC system — 16
   1.27 Principle of operation — 17

2 Gx-Developer – startup procedure — 18
   2.1 Opening a new project — 19
   2.2 Display settings – Zoom — 19
   2.3 Ladder diagram numbers — 22
   2.4 Project data list — 22

3 Producing a ladder diagram — 24
3.1 PLC program – FLASH1 — 24
3.2 Entering a ladder diagram — 25
3.3 Conversion to an instruction program — 27
3.4 Saving the project — 28
3.5 Program error check — 28
3.6 Instruction programming — 29
3.7 Program search — 31

4 Modifications to an existing project — 40
4.1 Copying a project — 40
4.2 Modification of the ladder diagram FLASH2 — 42
4.3 Modification details — 42
4.4 Deleting — 50

5 Serial transfer of programs — 55
5.1 Downloading a project to a PLC unit — 55
5.2 Executing the project — 56
5.3 Reducing the number of steps transferred to the PLC — 57
5.4 Communication setup — 58
5.5 System image — 59
5.6 Change of communications port — 60
5.7 Verification — 62
5.8 Uploading a project from a PLC — 63

6 Monitoring — 66
6.1 Ladder diagram monitoring — 66
6.2 Entry data monitoring — 67
6.3 Combined ladder and entry data monitoring — 70

7 Basic PLC programs — 71
7.1 Traffic light controller – TRAF1 — 71
7.2 Furnace temperature controller – FURN1 — 74
7.3 Interlock circuit – INTLK1 — 78
7.4 Latch relays — 80
7.5 Counters — 81
7.6 Online programming — 84
7.7 Batch counter – BATCH1 — 86
7.8 Assignment – BATCH2 — 87
7.9 Master control – MC1 — 87

8 PLC sequence controller — 91
8.1 Sequence function chart – SFC — 92
8.2 Ladder diagram – PNEU1 — 93
8.3 Simulation – PNEU1 — 97
8.4 Pneumatic panel operation — 98
8.5 Forced input/output — 101
8.6 Assignment – PNEU2 — 104

9 Free line drawing — 105
9.1 Inserting an output in parallel with an existing output — 108
9.2 Delete free line drawing — 109

10 Safety — 111
10.1 Emergency stop requirements — 111
10.2 Safety relay specification — 112
10.3 Emergency stop circuit – PNEU1 — 113
10.4 Safety relay – fault conditions — 114
10.5 System start-up check — 115

11 Documentation — 116
11.1 Comments — 117
11.2 Statements — 123
11.3 Display of comments and statements — 124
11.4 Comment display – 15/16 character format — 125
11.5 Comment display – 32 character format — 128
11.6 Notes — 130
11.7 Segment/note – block edit — 132
11.8 Ladder diagram search using statements — 133
11.9 Change of colour display — 134
11.10 Display of comments, statements and notes — 135
11.11 Printouts — 137
11.12 Multiple printing — 141
11.13 Saving comments in the PLC — 146

12 Entry ladder monitoring — 151
12.1 Ladder diagram – PNEU1 — 152
12.2 Principle of operation – entry ladder monitoring — 153
12.3 Deleting the entry ladder monitor diagram — 156

13 Converting a MEDOC project to Gx-Developer — 157
13.1 Importing a MEDOC file into Gx-Developer — 157

14 Change of PLC type — 162

15 Diagnostic fault finding — 165
15.1 CPU errors — 165
15.2 Battery error — 166
15.3 Program errors — 166
15.4 Help display – program errors — 168
15.5 Program error check — 169

16 Special M coils — 171
16.1 Device batch monitoring — 171
16.2 Option setup — 173
16.3 Monitoring the X inputs — 174

Contents

17 Set–reset programming — **175**
- 17.1 PNEU4 — 175
- 17.2 Sequence of operation – automatic cycle — 176
- 17.3 Sequence function chart – PNEU4 — 176
- 17.4 Ladder diagram – PNEU4 — 177
- 17.5 Principle of operation — 177
- 17.6 Simulation and monitoring procedure — 178
- 17.7 Monitoring PNEU4 — 179

18 Trace — **180**
- 18.1 Principle of operation — 180
- 18.2 Ladder diagram – PNEU4 — 181
- 18.3 Trace setup procedure — 182
- 18.4 Trace data — 182
- 18.5 Trace conditions — 183
- 18.6 Transfer Trace data to PLC — 185
- 18.7 Saving the Trace setup data — 185
- 18.8 Reading the Trace setup data from file — 186
- 18.9 Start Trace operation — 187
- 18.10 Start trigger – X0 — 189
- 18.11 Obtaining the Trace waveforms — 190
- 18.12 Trace results — 190
- 18.13 Measuring the time delay – T0 — 193
- 18.14 Calculation of elapsed time — 194

19 Data registers — **195**
- 19.1 Number representation – binary/decimal — 195
- 19.2 Converting a binary number to its decimal equivalent — 196
- 19.3 Binary numbers and binary coded decimal — 197
- 19.4 Advanced programming instructions — 198

20 Introduction to programs using data registers — **200**
- 20.1 Binary counter – COUNT3 — 200
- 20.2 BCD counter – COUNT4 — 202
- 20.3 Multiplication program – MATHS 1 — 205
- 20.4 RPM counter – REV1 — 206
- 20.5 Timing control of a bakery mixer – MIXER1 — 210

21 Ladder logic tester — **214**
- 21.1 Introduction — 214
- 21.2 Program execution — 214
- 21.3 Input simulation — 216
- 21.4 Device memory monitor — 217
- 21.5 Timing charts — 222
- 21.6 Producing the timing chart waveforms — 224
- 21.7 Resetting the timing chart display — 225

21.8	Saving the setup details	225
21.9	I/O system settings	225
21.10	Procedure – I/O system setting	226
21.11	Entering the Conditions and Input No. settings	228
21.12	Executing the I/O system	232
21.13	Resetting a data register using the I/O system	234
21.14	LLT2 modification	238
21.15	Simulating PNEU1 using ladder logic tester	240
21.16	PNEU1 procedure using ladder logic tester	241
21.17	Monitoring procedure	242

22 Bi-directional counters — 244

22.1	Ladder diagram – COUNT5	244
22.2	Special memory coils M8200–M8234	245
22.3	Principle of operation – COUNT5	245
22.4	Operating procedure	245
22.5	Monitoring – COUNT5	246

23 High-speed counters — 247

23.1	Introduction	247
23.2	Types of high-speed counters	247
23.3	FX range of high-speed counters	249
23.4	High-speed counter inputs	250
23.5	Up/down counting	251
23.6	Selecting the high-speed counter	251
23.7	Maximum total counting frequency	252
23.8	High-speed counter – HSC1	253
23.9	Decade divider – HSC2	254
23.10	Motor controller – HSC3	257
23.11	A/B phase counter – HSC4	260

24 Floating point numbers — 265

24.1	Floating point number range	265
24.2	Number representation	265
24.3	Floating point instructions	265
24.4	Storing floating point numbers – FLT1	266
24.5	Monitor – ladder diagram FLT1	267
24.6	Device batch monitoring	267
24.7	Floating point format	268
24.8	Obtaining the floating point value	269
24.9	Device batch monitoring – floating point numbers	270
24.10	Area of a circle – FLT2	270
24.11	Ladder diagram – FLT2	271
24.12	Principle of operation – FLT2	272
24.13	Monitored results – FLT2	273
24.14	Floating point – ladder logic tester	273

xii Contents

25 Master control – nesting **275**
    25.1 Nesting level 275
    25.2 Ladder diagram – MC2 276
    25.3 Principle of operation 277

26 Shift registers **279**
    26.1 Shift register applications 279
    26.2 Basic shift register operation 279
    26.3 Ladder diagram – SHIFT1 280
    26.4 Principle of operation – SHIFT1 280
    26.5 Operating procedure 281
    26.6 Monitoring – SHIFT1 281

27 Rotary indexing table **282**
    27.1 Index table system – plan view 282
    27.2 System requirements 283
    27.3 Shift register layout 284
    27.4 Ladder diagram – ROTARY1 285
    27.5 Principle of operation – ROTARY1 287
    27.6 Monitoring procedures 289
    27.7 Instruction scan and execution 291

28 Index registers V and Z **293**
    28.1 Index register instructions 293
    28.2 Stock control application – INDEX1 294
    28.3 System block diagram 294
    28.4 Warehouse – look-up table 294
    28.5 Ladder diagram – INDEX1 295
    28.6 Principle of operation 295
    28.7 Monitoring – INDEX1 296

29 Recipe application – BREW1 **298**
    29.1 System diagram 299
    29.2 Sequence of operation 299
    29.3 Recipe look-up tables 299
    29.4 Entering values into a look-up table (DWR) 300
    29.5 Downloading the recipe look-up table 303
    29.6 Selecting the device memory range 303
    29.7 Monitoring the recipe look-up table values 305
    29.8 Ladder diagram – BREW1 305
    29.9 Principle of operation – BREW1 306
    29.10 Monitoring – BREW1 307
    29.11 Test results 309
    29.12 Excel spreadsheet – recipe1 309

30 Sub-routines **310**
    30.1 Sub-routine program flow 311
    30.2 Principle of operation 311

	30.3	Temperature conversion – SUB1	312
	30.4	Ladder diagram – SUB1	312
	30.5	Labels	313
	30.6	Principle of operation – SUB1	313
	30.7	The sub-routine instructions	313
	30.8	Monitoring – SUB1	313
31	**Interrupts**		**315**
	31.1	Interrupt application	315
	31.2	Interrupt project – INT1	316
	31.3	Sequence of operation – automatic cycle	316
	31.4	Waveforms	317
	31.5	Ladder diagram – INT1	318
	31.6	Principle of operation – INT1	319
	31.7	Interrupt service routine	322
	31.8	Monitoring – INT1	322
32	**Step counter programming**		**324**
	32.1	Ladder diagram – STEP_CNTR1	325
	32.2	Principle of operation – STEP_CNTR1	326
	32.3	Simulation and monitoring procedure	329
	32.4	Entry data monitoring – STEP_CNTR1	330
	32.5	Pneumatic panel operation	331
33	**Automatic queuing system**		**332**
	33.1	System hardware	332
	33.2	FIFO memory stack	333
	33.3	Software diagram	333
	33.4	Ladder diagram – QUEUE1	334
	33.5	Principle of operation – QUEUE1	336
	33.6	Testing – QUEUE1	342
	33.7	Monitoring – QUEUE1	343
	33.8	Analysis of results	343
34	**Analogue to digital conversion FX2N-4AD**		**344**
	34.1	Introduction	344
	34.2	FX2N-4AD buffer memory addresses and assignments	344
	34.3	Voltage and current conversion	345
	34.4	Resolution – maximum input voltage	346
	34.5	Resolution – maximum input current	347
	34.6	Relationship between Vin and digital output	348
	34.7	ADC equations	349
	34.8	Resolution – independent of input voltage	350
	34.9	Highest possible resolution	351
	34.10	Example – voltage conversion	352
	34.11	Example – current conversion	353
	34.12	Count averaging	354
	34.13	Positioning the analogue unit	355

34.14	ADC wiring diagram	355
34.15	Hexadecimal numbering system for special units	356
34.16	Channel initialisation	356
34.17	TO and FROM instructions	357
34.18	ADC errors – BFM 29	359
34.19	Buffer memory – EEPROM	360
34.20	Software programming of offset and gain	360
34.21	Detecting an open circuit	361
34.22	Voltage/current specification	361
34.23	Ladder diagram – ADC1	362
34.24	Principle of operation – ADC1	364
34.25	Practical – analogue to digital conversion	366
34.26	ADC results	367
34.27	Monitoring using buffer memory batch	367
34.28	Test results	368

35 Digital to analogue conversion FX2N-4DA — 370

35.1	Introduction	370
35.2	Voltage resolution	370
35.3	FX2N-4DA buffer memory addresses and assignments	371
35.4	Error codes – BFM 29	372
35.5	Hardware diagram	373
35.6	DAC special unit no.1	373
35.7	Output mode select	374
35.8	Ladder diagram – DAC1	374
35.9	Principle of operation – DAC	375
35.10	Practical – digital to analogue conversion	377

36 Assignments — 378

Index — 387

Preface

This book comprehensively covers the programming and use of the complete range of Mitsubishi FX programmable controllers (PLCs), i.e. the FX1S, FX1N and the FX2N, unless otherwise stated.

For example the FX1S programmable controller cannot be used for:

1. Input/output expansion
2. Floating point arithmetic
3. Analogue to digital conversion
4. Digital to analogue conversion.

Since I wrote my first book on PLCs, *An Introduction to Programmable Logic Controllers*, which described the use of the DOS-based ladder diagram software MEDOC and the FX PLC, there have been enormous developments in computer technology in both hardware and software.

Mitsubishi Electric released the first version of their Windows-based programming software GPP Win in 1998 and at about the same time they produced the more powerful FXN range of PLCs.

The latest version of their Windows-based software is *Gx-Developer Version 8* on which this book is based and included with this software is the Gx-Simulator software known as *Ladder Logic Tester*.

The advantage of this simulator software is that it enables ladder diagram programs to be tested without the use of a PLC.

The Gx-Developer and the Gx-Simulator software can be used on Windows 98, 2000ME, 2000PRO and XP.

This book is intended for both students and engineers who wish to become competent in programming PLCs to meet the requirements of a wide variety of applications.

Students who are undertaking engineering courses will find the text covers most of the requirements for the following Edexcel Units.

1. BTEC National Certificate/Diploma in Electrical/Electronic Engineering Unit 31: Programmable Controllers.
2. BTEC Higher National Certificate/Diploma in Electrical/Electronic Engineering Unit 18: Programmable Logic Controllers.

GX Developer
A demo version of the GX Developer software from Mitsubishi Electric is available as a download from the companion site to this book:
http://books.elsevier.com/companions/0750656794

Acknowledgements

I have now been a PLC training engineer for over ten years and I have been most fortunate in meeting many PLC experts, who have willingly advised me on the capabilities of PLCs and the wide diversity of applications in which they have been used.

To them I say 'Many Thanks'

Andrew Brown	Senior Applications Engineer – Mitsubishi Customer Technology Centre.
Steve Case	Technical Support Engineer – Mitsubishi Customer Technology Centre.
David Dearden	UK Sales/Engineering Manager, Pilz Automation Technology, Corby.
Chris Evans	Manager, Mitsubishi Customer Technology Centre.
Dave Garner	Site Maintenance Manager, MFI Manufacturing, Runcorn.
Chris Garrett	Manager, Honda Assembly Support Department, Swindon.
Peter Hirstwood	Senior Project Engineer, Don Controls Ltd, Leeds.
Alex Kew	Applications Engineer – Mitsubishi Customer Technology Centre.
Roger Payne	Deputy Manager, Mitsubishi – Factory Automation Division, Hatfield.
Hinrich Schäfer	Managing Director, Schaefer Controls GmbH, Fuldatal Germany.
Jeremy Shinton	Software Business Development Manager, Mitsubishi UK.
Hugh Tasker	PLC Development Manager, Mitsubishi UK.
Jol Walthew	Manager, Mitsubishi Automation Training Department.

In addition, I would like to acknowledge with grateful thanks the support, encouragement and patience of my editors Rachel Hudson and Doris Funke at Elsevier, whilst writing this book.

Resources

The hardware/software used by the author to develop the programs used in this book is listed below.

Hardware	Supplier
FX2N 48MR-ES PLC	Mitsubishi Electric
FX2N-4AD – Analogue to digital converter	Mitsubishi Electric
FX2N 4DA – Digital to analogue converter	Mitsubishi Electric
8 input switch box	RS Components
4 digit thumbwheel switch	RS Components
4 digit BCD display	London Electronics Ltd, Chicksands, Bedfordshire
Pneumatic Panel	SMC Pneumatic(UK) Ltd, Milton Keynes, Northamptonshire
Safety Relay	Pilz Automation, Corby, Northamptonshire

Software	Supplier
Gx Developer Ver. 8.03	Mitsubishi Electric
Gx Simulator Ver. 6.13	Mitsubishi Electric

Resources

Glossary

ADC Analogue to digital converter.
Boolean algebra A mathematical method for simplifying logic gate circuits and PLC ladder diagrams.
Copying Used to make a copy of a ladder diagram but having a different filename.
COM1 Communications port no.1. A serial communications port, which is usually a 9-pin plug situated at the rear of a computer. It enables communications between the computer and the PLC.
CPU Central processor unit. The CPU is the microchip within the PLC, which executes the Instruction Program obtained from the converted ladder diagram.
DAC Digital to analogue converter.
DOS Disk operating system. A Microsoft computer operating system, which was a forerunner to Windows.
Downloading Serial transfer of the Instruction Program from the computer to the PLC, via a COM or USB port.
Find A Gx-Developer function used for finding a device, an instruction or the start of a required ladder diagram line.
Gx-Developer The Mitsubishi Electric ladder diagram software, whose use is described in this book.
Gx-Simulator This Mitsubishi Electric software enables ladder diagram programs to be tested without the need for a PLC. Also known as Logic Ladder Tester.
Instruction When used with a Mitsubishi Electric FX2N, there are approximately 160 instructions, which range from the simple MOVE instruction to complex mathematical instructions.
Instruction Program An alternative method to a ladder diagram, for producing a PLC program. It is similar to a low-level assembly type microprocessor program. It was, at one time, the only method available for producing PLC programs before the availability of ladder diagram software. All ladder diagram programs though have to be converted to Instruction Programs for downloading to the PLC. It is not possible for ladder diagram programs to be directly downloaded or uploaded to/from a PLC.
I/O This term is used to describe the total number of inputs and outputs allocated to a particular PLC. For example, an FX2N 48 MR PLC has 24 X inputs and 24 Y outputs.
Karnaugh Map A pictorial method for simplifying logic gate circuits and PLC ladder diagrams.
MEDOC A DOS-based Mitsubishi Electric software package for producing ladder diagrams, the forerunner to Gx-Developer.
PCMCIA An abbreviation for Personal Computer Memory Card International Association. The PCMCIA card can slot into a laptop computer or some PLCs.

It was initially designed for adding memory to laptop computers but can also be used to provide extra memory to PLCs, i.e. the Mitsubishi Q2AS. A PCMCIA card is also available as an interface for an RS232 connection with laptop computers, which do not have this type of serial output connection.

PLC An abbreviation for Programmable Logic Controller. It is used in the automatic control of machinery and plant equipment. The great advantage of a PLC is that it can be programmed using software, i.e. Gx-Developer, to carry out a wide variety of tasks.

RS232 A serial communications system used for transferring information. Used in a PLC system for downloading and uploading instruction programs.

Safety Relay A Safety Relay is used in control systems to ensure that that the system will fail safe whenever an Emergency Stop condition occurs.

SC09 A Mitsubishi RS232 serial communications cable. It connects the computer to the PLC to enable programs to be downloaded, uploaded and monitored.

Sink A term, which describes the direction of current flow into or out of either an input or an output terminal of a PLC. To operate a PLC sink input, the direction of current flow will be out of the PLC X input terminal through a closed input switch/proximity detector and then into a 0 V terminal.

Source A term, which describes the direction of current flow into or out of either an input or an output terminal of a PLC. To operate a PLC source input, the direction of current flow will be from a positive voltage supply through a closed input switch/proximity detector and then into the PLC input terminal. The FX2N range of PLCs includes a transistor source output type, which means that current will flow out the Y output terminal through the output load and to a 0 V connection.

USB An abbreviation for Universal Serial Bus. A modern type of communications system for connecting peripherals, i.e. mouse, printer, scanner, internet modem, to a computer. There are adapters available, which enable a Mitsubishi Electric SC09 communication cable to be connected to the USB port of a computer.

1
Introduction to PLCs

The need for low-cost, versatile and easily commissioned controllers has resulted in the development of programmable logic controllers, which can be used quickly and simply in a wide variety of industrial applications.

The most powerful facility which PLCs have, is that they can be easily programmed to produce their control function, instead of having to be laboriously hard-wired, as is required in relay control systems.

However, the method of programming a PLC control system can nevertheless use relay ladder diagram techniques, which therefore enables the skills of an outdated technology to be still viable with that of the new.

The PLC was initially designed by General Motors of America in 1968, who were interested in producing a control system for their assembly plants and which did not have to be replaced every time a new model of car was manufactured.

The initial specification for the PLC was:

1. Easily programmed and reprogrammed, preferably in plant, to enable its sequence of operations to be altered.
2. Easily maintained and repaired.
3. More reliable in a plant environment.
4. Smaller than its relay equivalent.
5. Cost-effective in comparison with solid-state and relay systems, then in use.

1.1 Basic PLC units

The four basic units within the FX2N PLC units are:

1. The central processor unit (CPU)
This is the main control unit for the PLC system, which carries out the following:
    (a) Downloads and uploads ladder diagram programs via a serial communications link.
    (b) Stores and executes the downloaded program.
    (c) Monitors in real time the operation of the ladder diagram program. This gives the impression that a real hardwired electrical control system is being monitored.
    (d) Interfaces with the other units in the PLC system.

2   *Mitsubishi FX Programmable Logic Controllers*

2. Input unit
   The input unit enables external input signals, i.e. signals from switches, push buttons, limit switches, proximity detectors, to be connected to the PLC System and then be processed by the CPU.
3. Output unit
   The output unit is connected to its externally operated devices, i.e. LED's, indicator lamps, digital display units, small powered relays, pneumatic/hydraulic pilot valves.

   Each time the program is executed, i.e. after each program scan, then depending on the ladder diagram program and the logic state of the inputs, the outputs will be required to turn ON, turn OFF, or remain as they are.
4. Power supply
   The power supply is used to provide the following DC voltages from the 240 V mains supply:
   (a) 5 V DC supply for the internal electronics within all of the PLC units.
   (b) 24 V DC supply, which can be used to supply the input devices.
   (c) Alternatively, the 24 V DC can be supplied from an external DC power supply, which is used for both the input and the output devices.

1.2 Comparison of PLC and RELAY systems

Characteristic	PLC	Relay
Price per function	Low	Low – if equivalent relay program uses more than ten relays
Physical size	Very compact	Bulky
Operating speed	Fast	Slow
Electrical noise immunity	Good	Excellent
Construction	Easy to program	Wiring – time-consuming
Advanced instructions	Yes	No
Changing the control sequence	Very simple	Very difficult – requires changes to wiring
Maintenance	Excellent – PLCs rarely fail	Poor – relays require constant maintenance

1.3 PLC software

To be able to design a PLC program using a computer, it is essential for the software to have the following facilities:

1. Programs can be designed using conventional relay ladder diagram techniques.
2. Test if the program is valid for use on the chosen PLC.
3. Programs can be permanently saved either on a computer's hard disk or on floppy disks.
4. Programs can be re-loaded from either the hard disk or the floppy disk.
5. Ladder diagram contacts and coils can be annotated with suitable comments.
6. Hard copy printouts can be obtained.
7. The program can be transferred to the PLC, via a serial link.
8. The program within the PLC can be transferred back to the computer.

9. The ladder diagram control system can be monitored in 'real time'.
10. Modifications can take place, whilst the PLC is online.

1.4 Gx-Developer software

The Gx-Developer software is a Windows-based package, which enables users to produce ladder diagram projects for use with the Mitsubishi range of PLCs.

It has been produced by Mitsubishi Electric to replace the DOS-based package, MEDOC.

Advantages – Gx-Developer

1. As the software uses drop-down menus, there is no need to remember keypress characters.
2. The drop-down menus are selected using a mouse.
3. All of the functions can be accessed using an icon, instead of the drop-down menus.
4. Ladder diagrams can be entered more quickly.
5. Modifications can be easily carried out.
6. Improved monitoring facilities, i.e. direct monitoring of the contents of a special unit's buffer memory.
7. Fault-finding diagnostics.
8. Improved documentation, i.e. notes.

1.5 Hardware configuration

This section deals with configuring an FX2N system.

Since the main components of all FX PLCs, i.e. the CPU, inputs and outputs are all parts of the one unit instead of separate plug in modules, the FX range of PLCs are known as 'Brick Type' PLCs.

The main considerations that must be taken into account when configuring a system are:

1. External devices, inputs and outputs.
   (a) How many are required?
   (b) Is the supply from the Input devices to the PLC inputs from: volt-free contacts, 24 V DC, or 110 V AC?
   (c) Is the supply from the PLC outputs to the external loads from: volt-free contacts, 24 V DC, or 110 V AC?
   (d) Is a fast-switching operation required?
   (e) Are proximity detectors required (see Section 1.21)?
2. Power supply requirements.
   (a) Supply voltage.
   (b) Internal power supply.
3. Special function units.
   (a) How many can the system support?
   (b) Is an external power supply required?

1.6 Base unit, extension units and extension blocks

Figure 1.1 shows a base unit along with 2 × extension blocks.

It is very important that confusion is avoided when these units are discussed.

The basic way to describe the difference between a base unit, an extension unit and an extension block is as follows:

1. A base unit is made up of four components, i.e. power supply, inputs, outputs and CPU.
2. An extension unit is made up of three components, i.e. power supply, inputs and outputs.
3. An extension block is made up of one or two components, i.e. inputs and/or outputs. It can be seen that the extension block does not have a power supply. It therefore obtains its power requirement from either the base unit or an extension unit. Hence it is necessary to determine how many of these un-powered units can be connected before the 'On Board' power supply capacity is exceeded. The tables in Sections 1.8 and 1.9 show how this can be worked out.

1.7 PLC voltage supplies

24 V DC supply

The FX2N has a 24 V internal power supply, which can be used for supplying current to input switches and sensors.

From the smaller FX2N table below, it can be seen that if no un-powered extension blocks have been used, then the maximum available current from the 24 V supply is 250 mA.

However, if one 16-input and one 16-output extension block were fitted, then the available current falls to 0 mA and a separate 24 V power supply would then be required for supplying the input switches and any sensors.

1.8 Smaller FX2N PLCs

A = Number of additional outputs   FX2N-16M* - E** → FX2N-32M* - E**
B = Number of additional inputs   FX2N 32E* -E**
C = Invalid configuration

Available current (mA)

A		B=0	B=8	B=16	B=24	B=32
A	24	25		C		
	16	100	50	0		
	8	175	125	75	25	
	0	250	200	150	100	50

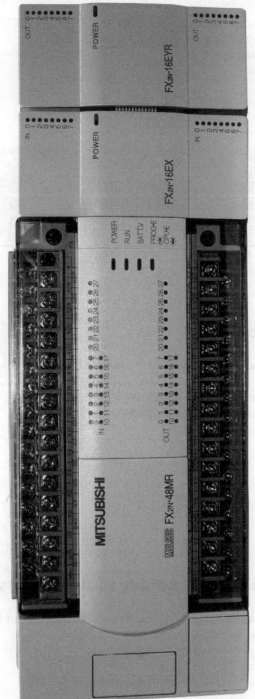

Figure 1.1

1.9 Larger FX2N PLCs

A = Number of additional outputs   FX2N 48M* - E** → FX2N-128M* - E**
B = Number of additional inputs   FX2N 48E* - E**
C = Invalid configuration

Available current (mA)

A \ B	0	8	16	24	32	40	48	56	64
48	10								
40	85	35							
32	160	110	60	10		C			
24	235	185	135	85	35				
16	310	260	210	160	110	60	60		
8	385	335	285	235	185	185	135	35	
0	460	410	360	310	260	210	160	110	60

1.10 5 V DC supply

The FX2N has a second power supply, of 5 V, which is not available to the user.

Its function is to supply, via the ribbon cable bus connections, any special units connected to the system.

The table below details the current available from this supply.

Unit	Max. 5 V DC bus supply
FX2N- **M*- ES (ESS)	290 mA
FX2N - **E * - ES (ESS)	690 mA

1.11 Special unit power supply requirements

Depending on the special units used, the current consumption from the 5 V supply and the 24 V supply must be taken into account.

The table on page 7 gives the current required by the most frequently used units along with the I/O requirements.

Introduction to PLCs 7

Model	Description	No. of I/O	Supply from PLC 5 V bus	24 V supply current
FX2N-4AD	Analog to digital converter	8	30 mA	200 mA
FX2N-4DA	Digital to analog converter	8	30 mA	55 mA
FX2N-4AD-PT	PT100 probe interface	8	30 mA	50 mA
FX2N-4AD-TC	Thermocouple interface	8	30 mA	50 mA
FX2N-1HC	High-speed counter	8	90 mA	–
FX2N-1PG	Pulse output position control	8	55 mA	40 mA

1.12 Part number

The part number describes the type of PLC and its functionality.
The part number can be broken down as in Figure 1.2

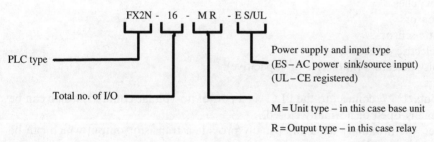

Figure 1.2

1.13 Serial number

Also found on the unit is a serial number (Figure 1.3), from which the construction date can be determined.

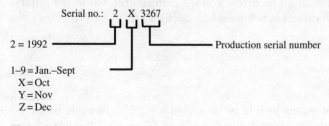

Figure 1.3

8 Mitsubishi FX Programmable Logic Controllers

1.14 PLC inputs

All PLC inputs are isolated by photocouplers to prevent operational errors due to contact chattering or other electrical noise that may enter via the input.

For this very reason ON/OFF status changes will take approximately 10 msec. This time should be taken into account when programming, as it will have a direct effect on the way the program will operate.

For the input device to actually register on the PLC it will have to draw a minimum of 4 mA for the PLC input to switch. Anything less than 4 mA, will result in the PLC input not turning on.

The current into a PLC input must not exceed 7 mA; anything in excess of this could result in the input being damaged.

The input signals can come from a wide variety of devices, i.e.

1. Push buttons.
2. Rotary switches.
3. Key switches.
4. Limit switches.
5. Level sensors.
6. Flow rate sensors.
7. Photo-electric detectors.
8. Proximity detectors (inductive or capacitive).

The inputs '1'–'7' connect to the PLC via a pair of no-voltage contacts, which can be either normally open or normally closed.

However, the proximity detectors usually provide a transistor output which can be either an NPN or a PNP transistor.

1.15 AC inputs

110 V AC Inputs are also available.

It is recommended that the same supply voltage to the PLC is used as for the inputs, i.e.(100–120 V AC).

This minimises the possibility of an incorrect voltage being connected to the inputs. With AC versions the S/S terminal is not used (see Section 1.18).

Note

1. In normal operation, use of inputs should be restricted to 70% at any one time.
2. Except for inputs concerned with safety (refer page 76 and Chapter 10) input devices such as ON/OFF switches, push buttons, foot switches and limit switches are usually wired to the PLC through the normally open contacts of the device.

1.16 PLC outputs

There are three different types of output for the FX range of PLCs, these are:

1. Relay.
2. Triac (solid-state relay – SSR).
3. Transistor.

Relay

This is the most commonly used type of output.

The coils and the contacts of the output relays enable electrical isolation to be obtained between the internal PLC circuitry and the external output circuitry.

Dependent on a number of factors, i.e. the supply voltage, the type of load, i.e. resistive, inductive or lamp, the contact life, the maximum-switched current per individual output is 2 A.

The PLC will provide groups of 4, 8 or 16 outputs each with a common. The commons are logically numbered COM1, COM2, etc. and are electrically isolated from one another.

When the 'END' instruction in the ladder diagram is executed, the PLC will REFRESH the outputs from the output latch memory to turn the appropriate output relay either ON or OFF.

The response time for the operation of an output relay is approximately 10 msec.

Triac

The TRIAC is an AC switch, which basically consists of two thyristors connected 'back to back'.

Since the TRIAC output is solid state, the lifetime of a TRIAC output is far longer than that of the relay output.

The voltage range of these devices is 85–240 V AC and each output can switch up to a maximum of 0.35 A.

As with all other output configurations, the physical output is isolated by a photocoupler.

The response of the TRIAC when turning ON is faster than the Relay, i.e. 1 msec but the OFF times are identical, i.e. 10 msec.

Care should be taken when configuring the system so that the output circuitry is not overloaded.

Care should also be taken concerning leakage current in a TRIAC output circuit. This current is far greater than that of a relay circuit and may cause any externally connected miniature relays to remain energised.

Transistor

The transistor outputs are used, where a very fast switching time is required.

The switching time of the transistor outputs, whether they are Sink or Source outputs, is < 0.2 msec with a 24 V DC, 100 mA load.

As with all other output configurations, the physical output is isolated by a photocoupler.

1.17 Source–sink inputs

The term source–sink refers to the direction of current flow into or out of the input terminals of the PLC.

Source input

When the PLC is connected for source inputs, then the input signal current flows into the X inputs (Figure 1.4).

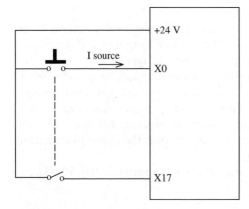

Figure 1.4

Sink input

When the PLC is connected for Sink inputs, then the input signal current flows out of the X inputs (Figure 1.5).

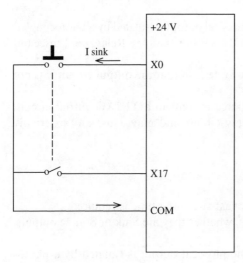

Figure 1.5

1.18 The source/sink – S/S connection

The S/S connection is the common terminal for all of the internal input circuits of the PLC.

It enables the user to decide the direction in which the input devices will supply current to the PLC inputs, i.e. source or sink.

1.19 Source inputs – block diagram

To ensure that all of the input devices will supply the source input current, the user connects the S/S terminal to the 0 V terminal, as shown in Figure 1.6.

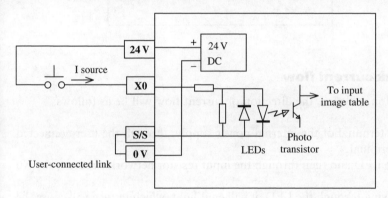

Figure 1.6

Direction of source current flow

When the push button is closed, the direction of current flow will be as follows:

1. From the +24 V terminal of the internal power supply, the +24 V PLC terminal, and then through the push button and into the X0 input terminal, i.e. source current.
2. Through the input resistor network circuit and then through the second LED.
3. With current flowing through the LED it will emit light, which in turn will cause the photo-transistor to turn ON.
4. The function of the photo-transistor is to isolate the 24 V input circuit from the 5 V PLC logic circuit and hence increase the noise immunity of the input.
5. With the photo-transistor turning ON, this will cause a signal to be sent to the input image table, to store the information that the input X0 is ON.
6. The input current now flows to the S/S terminal, through the user-connected link to the PLC 0 V terminal and then back to the negative (−) terminal of the internal power supply.

1.20 Sink inputs – block diagram

To ensure that all of the input devices will sink the current from the PLC inputs, the user now connects the S/S terminal to the +24 V terminal, as shown in Figure 1.7.

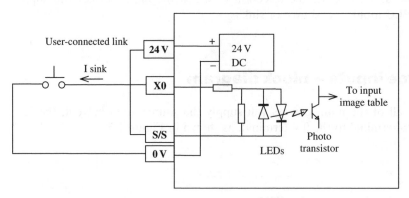

Figure 1.7

Direction of sink current flow

When the push button is closed, the direction of current flow will be as follows:

1. From the +24 V terminal of the internal power supply, through the user-connected link to the S/S terminal.
2. Through the first LED and then through the input resistor network circuit to the X0 input terminal.
3. With current flowing through the LED, it will emit light, which in turn will cause the same photo-transistor to turn ON.
4. With the photo-transistor turning ON, this will cause a signal to be sent to the input image table, to store the information that the input X0 is ON.
5. The input current now flows out of the X0 input terminal, i.e. sink current.
6. It then flows through the push button to the PLC 0 V terminal and then back to the negative terminal of the internal power supply.

1.21 Proximity sensors

There are two types of proximity sensors i.e. inductive and capacitive, and the supply voltages to these sensors are normally 24 V DC.

There are also two standard outputs for both proximity sensors, which are:

1. PNP (source)
2. NPN (sink)

Once selected, only that output type can be used for supplying the inputs to the PLC. They *cannot* be mixed.

If PNP proximity detectors are used, then every one of the PLC inputs become source inputs.

If NPN proximity detectors are used, then every one of the PLC inputs become sink inputs.

To configure the PLC to accept either a PNP or an NPN sensor, the S/S terminal has to be linked to either the 0 V line or the 24 V DC line respectively, as shown in the Figure 1.8.

Care must be taken to ensure that the S/S terminal is correctly connected, as failure to do this will result in the input not working.

1.22 S/S terminal configurations

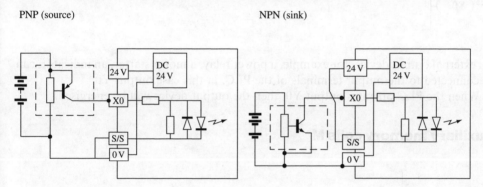

Figure 1.8

1.23 PLC ladder diagram symbols

Inputs X

Normally open contact

When an external source, e.g. an external switch, push button, relay contact, etc., operates, then the corresponding ladder diagram normally open contact or contacts, will close.

The X1 indicates that the external input is connected to input X1 of the PLC.

Normally closed contact

```
 X2
─┤/├─
```

When the external input connected to the PLC is operated, then the corresponding ladder diagram contact or contacts will open.

Outputs Y

```
──( Y0 )┤
```

An external output device, for example, a power relay, a motor starter, an indicator, can be connected to the output terminals of the PLC, in this case output Y0.

When the PLC operates output Y0, then the output device will be energised.

Auxiliary memory coils M

```
─(M0    )┤
```

An Auxiliary Memory Coil can be used in PLC programs for a variety of reasons.

1. To operate when the set of inputs, which are connected to the M Coil, are correct.

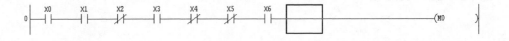

The inputs corresponding to the normally open contacts have been operated, i.e. X0, X1, X3, X6. The inputs corresponding to the normally closed contacts have not been operated, i.e. X2, X4, X5. This information can then be used throughout the ladder diagram by simply using the contacts of the memory coil, i.e. M0 instead of having to repeat all of those input contacts, which caused the M coil to initially operate.
2. As part of a latch circuit.
3. As part of a shift register circuit.

1.24 PLC address ranges

The following range of addresses are those used for the FX2N 48 I/O base unit.

Inputs
X0–X27 (octal) 24 inputs.
Expandable inputs 4–24.

Outputs
Y0–Y27 (octal) 24 outputs.
Expandable outputs 4–24.

Timers

T0–T199	0.1 sec–3276.7 sec
T200–T245	0.01 sec–327.67 sec
T246–T249	0.001 sec–32.767 sec retentive and battery-backed.
T250–T255	0.01 sec–3276.7 sec retentive and battery-backed.

Counters

C0–C99	general-purpose (16 bit)
C100–C199	battery-backed (latched 16 bit)
C200–C219	bi-directional (32 bit)
C220–C234	bi-directional and battery-backed
C235–C255	high-speed counters

Auxiliary relays

M0–M499	general-purpose
M500–M3071	battery-backed
M8000–M8255	special-purpose

State relays

S0–S999	general-purpose
S500–S999	battery-backed
S900–S999	annunciator

Data registers

D0–D199	general-purpose
D1000–D7999	file registers
Selectable from battery backup range	
D200–D7999	battery-backed
D8000–D8255	special-purpose
V and Z	index registers V0–V7 and Z0–Z7 (16 bit)

1.25 Basic operation of a PLC system

To explain the basic operation of a PLC system, consider the following two lines of program:

```
     X1
0 ───┤ ├──────────────────────────────────( M0 )─

     M0
2 ───┤ ├──────────────────────────────────( Y1 )─

4 ────────────────────────────────────────[ END ]─
```

16 *Mitsubishi FX Programmable Logic Controllers*

1. When Input X1 closes, this operates internal memory coil M0.
2. The normally open contact of M0 on closing will cause output Y1 to become energised.

1.26 Block diagram – basic operation of a PLC system

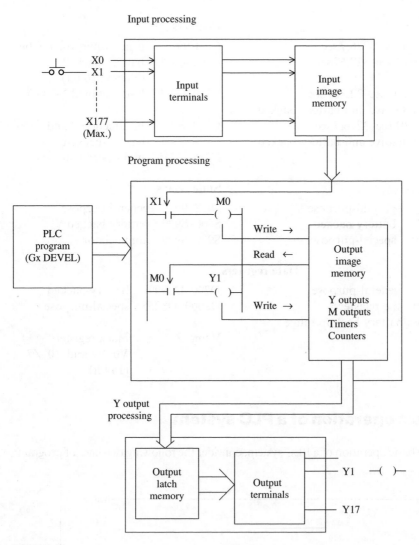

Figure 1.9

1.27 Principle of operation

Input processing

The PLC initially reads the ON/OFF condition of all of the inputs used in the program. These conditions are then stored into the input image memory.

Program processing

1. The PLC then starts at the beginning of the PLC program, and for each element of the program, it READS the actual logic state of that element, which is stored in either the input image memory or the output image memory.
2. If the required logic state is correct, i.e. X1 is ON, the PLC will move on to the next element in the rung, i.e. M0.
3. If X1 is ON, then a logic 1 will be WRITTEN into the output image memory in the location reserved for M0.
4. If X1 is OFF, then a logic 0 is WRITTEN into the M0 memory location.
5. After an output instruction has been processed, the first element on the next line is executed, which in this example is a normally open contact of M0.
6. Hence the logic state of the M0 memory location is this time READ from, and if its logic state is at logic 1 indicating that the M0 coil is energised, this effectively means all M0 normally open contacts will now close. The contact of M0 being closed, will cause a Logic 1 to be WRITTEN to the memory location reserved for the output Y1.
7. However, if the contents of the M0 memory location are at logic 0, i.e. M0 is not energised, then a Logic 0 is WRITTEN to the Y1 memory location.

Output processing

1. Upon completion of the execution of all instructions, the contents of the Y memory locations within the output image memory are now transferred to the output latch memory and the output terminals.
2. Hence, any output, which is designated to be ON, i.e. Y1, will become energised.

2
Gx-Developer – startup procedure

1. Ensure the Gx-Developer software has been installed in the computer.
2. From Windows desktop, select the Gx-Developer icon.
3. The display now becomes, as shown in Figure 2.1.

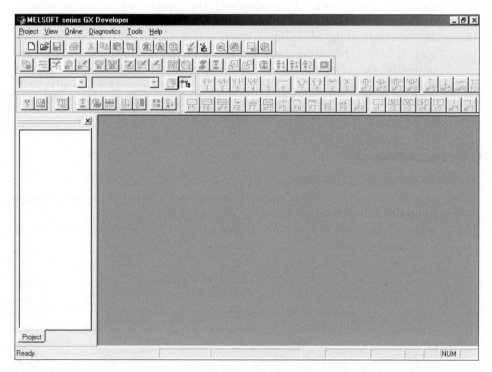

Figure 2.1

4. As can be seen from Figure 2.1 there are large number of icons and this can be confusing to the first-time user. Hence, initially, only an essential minimum number of icons will be displayed.
5. From the main menu, select View and then Toolbar.

6. Delete the items, which no longer are identified by an X, so that the display appears as shown in Figure 2.2.

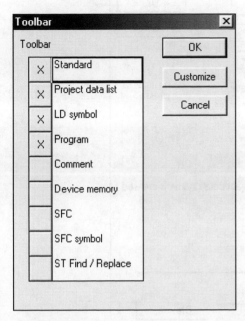

Figure 2.2

7. Select OK, to return to the Main Display.

2.1 Opening a new project

1. From the main menu, select Project.
2. Select New Project.
3. Enter the details as shown in Figure 2.3, i.e.
    (a) PLC Series   FXCPU.
    (b) PLC Type   FX2N(C).
    (c) Setup project name   ✓.
    (d) Drive/Path   c:\gxdevel\fxi.
    (e) Project name   flash1.
    (f) Select OK.
    (g) Select Yes.
4. The display now becomes as shown in Figure 2.4.

2.2 Display settings – Zoom

To obtain the ladder diagram display, within the width of the VDU screen, it is first of all necessary to adjust the Zoom settings.

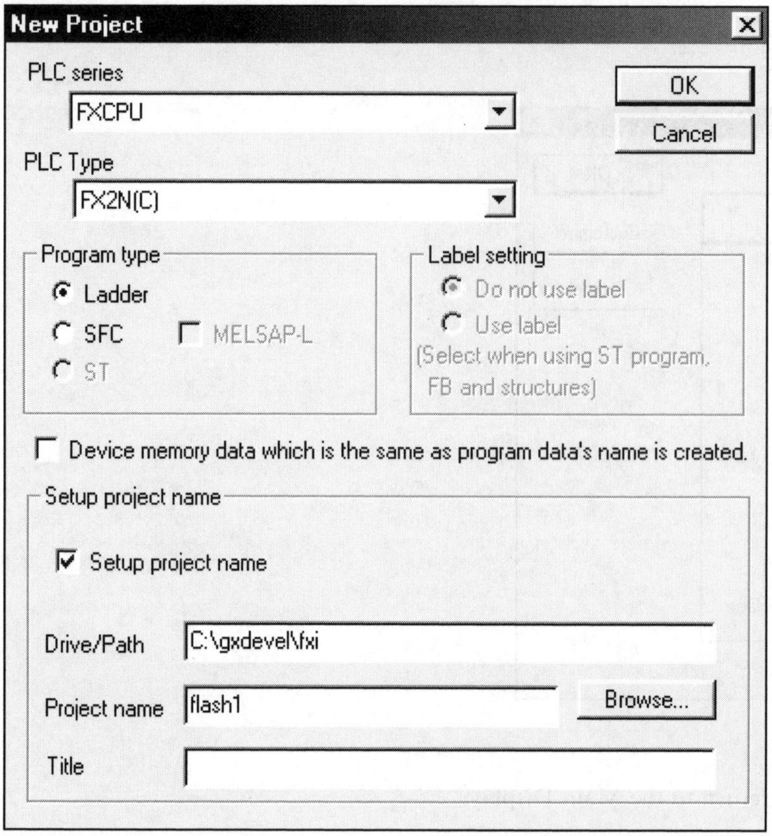

20 *Mitsubishi FX Programmable Logic Controllers*

Figure 2.3

This is done as follows:

1. From the main menu, select View.
2. Select Zoom.
3. Select Auto.
4. This ensures that the ladder diagram will be displayed within the width of the VDU screen.
5. The ladder diagram now becomes as shown in Figure 2.5.
6. Note the following:
   (a) The main program is FLASH1.
   (b) The project data list is shown on the left-hand side of the display.
   (c) The project parameters can be selected and viewed from the project data list.
   (d) The ladder diagram is displaying the final line, i.e. END.
   (e) The ladder diagram symbols are now highlighted.

Gx-Developer – startup procedure 21

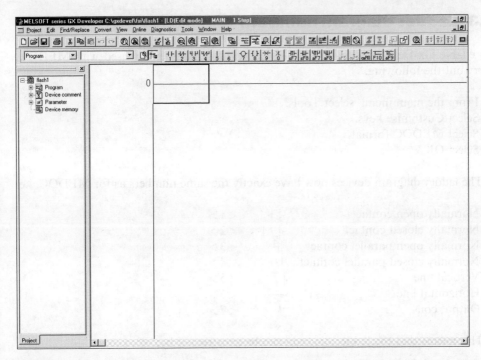

Figure 2.4

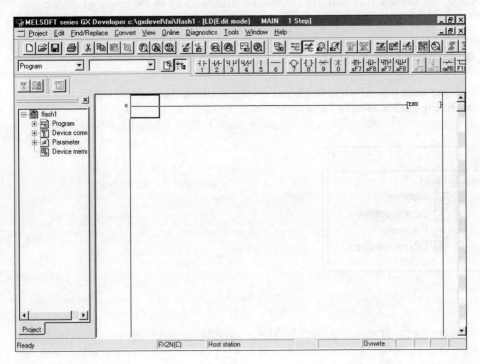

Figure 2.5

2.3 Ladder diagram numbers

To enable Gx-Developer to have the same keypress numbers as used with MEDOC, carry out the following:

1. From the main menu, select Tools.
2. Select Customise keys.
3. Select MEDOC format.
4. Select OK.

The ladder diagram devices now have exactly the same numbers as for MEDOC, i.e.

1. Normally open contact   -| |-   <1>
2. Normally closed contact   -| / |-   <2>
3. Normally open parallel contact   ⊔|⊔   <3>
4. Normally closed parallel contact ⊔/⊔   <4>
5. Vertical line | <5>
6. Horizontal line   —   <6>
7. Output coil   -( )-   <7>

This means that the ladder diagram can be constructed by:

1. Using the mouse and selecting the required device.
2. Using the keyboard to enter the number corresponding to the required device.

2.4 Project data list

The project data list, which is displayed on the left-hand side of the ladder diagram as shown in Figure 2.6, is used for a variety of purposes, i.e.

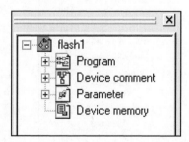

Figure 2.6

1. Rename the program name from MAIN to something more appropriate. This is used especially with the Q Series PLCs, which have the facility to store more than one program at a time.

2. Enable comments to be saved in the PLC (see Section 11.13).
3. Enter numerical values and download them directly into Device Memory (see Section 29.4).

However, at this point there is no need for the project data list and therefore it can be removed by clicking on the ☒ in the top right-hand corner of the project data list.

Project data list icon

Alternatively, the project data list can be toggled ON/OFF by selecting the project data list icon (Figure 2.7).

Figure 2.7

3

Producing a ladder diagram

3.1 PLC program – FLASH1

The program FLASH1 enables a PLC output, i.e. Y0, to be turned ON/OFF at a controlled rate.

In this example, the Output Y0 will be ON for 1 sec and then OFF for 1 sec.

It will be used to describe how a PLC ladder can be produced, modified and tested. Then using a Mitsubishi FX2N PLC, the program will be downloaded, run and monitored.

PLC ladder diagram – FLASH1

```
         X0      T1                                      K10
0       ─┤ ├────┤/├──────────────────────────────────( T0    )─

         T0                                              K10
5       ─┤ ├──────────────────────────────────────────( T1    )─

         T0
9       ─┤ ├──────────────────────────────────────────( Y0    )─

11      ───────────────────────────────────────────────[ END  ]─
```

Line numbers

In the descriptions that follow, references will be made to line numbers.

A line number is the step number of the first element for that particular line.

Therefore, line numbers will not increase by one from one line to the next, but will depend on the number of steps used by the elements, for each line.

Principle of operation

1. Line 0
   (a) On closing the input switch X0, the timer T0 will be enabled via the normally closed contact of timer T1.

(b) Timer T0 will now start timing out, and after 1 sec, the Timer will operate. This means:
   (i) Any T0 normally open contacts -||-, will close.
   (ii) Any T0 normally closed contacts -|/|-, will open.
2. Lines 5 and 9
   There are two T0 contacts, which are both normally open, therefore both of them will close, causing the following to occur:
   (a) Timer T1 will become enabled and start timing out.
   (b) Output Y0 will become energised, i.e. output Y0 will turn ON.
3. Line 5
   After timer T1 has been energised for 1 sec, it will also operate and its normally closed contact will open, causing Timer T0 to dropout.
4. With Timer T0 dropping-out, its normally open contact will now re-open causing:
   (a) Timer T1 to dropout even though T1 has just timed out.
   (b) Output Y0 to become de-energised, i.e. output Y0 will turn OFF.
5. Hence it can be seen that timer T1 is part of a 'cut-throat' circuit in that when it does time out, it immediately de-energises itself.
6. With timer T1 dropping-out, its normally closed contact will close, and for as long as input X0 is closed, the operation will be constantly repeated.
7. Line 9
   Hence the output Y0 will be continuously OFF for 1 sec and then ON for 1 sec.

3.2 Entering a ladder diagram

The ladder diagram of FLASH1, as shown on page 24, will now be entered.

1. Entering the first contact, normally open X0.
   (a) Using the mouse select the normally open contact (Figure 3.1).

Figure 3.1

   (b) Enter the device name X0.
   (c) Select OK.
   (d) The ladder diagram now becomes as shown below:

26 *Mitsubishi FX Programmable Logic Controllers*

2. Second contact – normally closed T1.
   Use the keyboard to enter the following:
   (a) 2 for a normally closed contact.
   (b) T1.
   (c) Enter.
3. The ladder diagram now becomes as shown below:

4. Output, timer T0.
   Enter the following (Figure 3.2):
   (a) 7 for a coil.
   (b) T0.
   (c) Space.
   (d) K10.
   (e) Enter.

Figure 3.2

5. Note:
   Unlike MEDOC, a space and not <Enter> is used between the timer T0 and its time delay value K10.
6. The first line of the ladder diagram is now as shown in Figure 3.3.

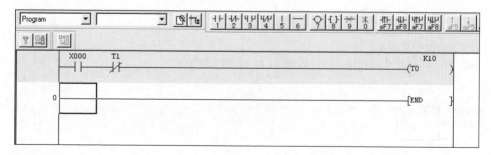

Figure 3.3

7. Complete the remainder of the ladder diagram, as shown on page 24 by entering the following:
   (a) 1 T0 <ent>.
   (b) 7 T1 <space> K10 <ent>.
   (c) 1 T0 <ent>.
   (d) 7 Y0 <ent>.

END instruction

There is no need to enter the instruction END as it is always on the last line of the ladder diagram.

Complete ladder diagram – FLASH1

3.3 Conversion to an instruction program

1. Before the program can be saved, the ladder diagram must first of all be converted into a set of instructions.
2. To execute the conversion process, carry out the following:
   (a) From the main menu, select Convert.
   (b) From the Convert menu, select Convert F4.
3. The display now becomes as shown below:

4. Note:
   (a) The grey unconverted background area becomes clear.
   (b) Line numbers appear at the start of each line.

Conversion process – F4

For the remainder of the course, the function key F4 will be used for converting the ladder diagram to its equivalent instruction program.

3.4 Saving the project

To save the project on the hard drive, carry out the following:

1. From the main menu, select Project.
2. Select Save.
3. The project FLASH1, will now be saved in the folder c:\gxdevel\fxi\flash1.
4. Alternatively, from the main toolbar, select the save icon (Figure 3.4).

Figure 3.4

3.5 Program error check

After a ladder diagram has been produced, it is advisable to check that it does not contain any errors.

The types of errors, which are checked for are:

1. The correct instructions have been used for the PLC type.
2. The same output coil, i.e. Y0 has not been used more than once.
3. There is consistency concerning paired instructions, i.e. SET and RESET, MC and MCR.

   1. From the Tools menu, select Check program (Figure 3.5).
   2. Select Execute.

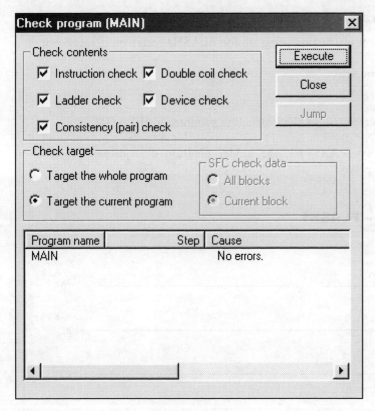

Figure 3.5

3. If there are no errors, then the message No errors is displayed.

3.6 Instruction programming

Once a ladder diagram has been produced it has to be converted to an instruction program as only the instruction program can be saved and downloaded to the PLC.

However, when using Gx-Developer it is possible to write an instruction program directly but, unless a programmer is very skilled at producing such programs, it is very unlikely that a complex program can be produced using this method.

Prior to the availability of ladder diagram software, a PLC program could only be produced using an instruction program.

Where Gx-Developer has been used to produce a ladder diagram, then the equivalent instruction program can easily be displayed.

Instruction program – FLASH1

To obtain the equivalent instruction program for FLASH1, carry out the following:

1. From the main menu select:
   (a) View.
   (b) Instruction list.
2. Displayed on the screen will be the instruction program for FLASH1.
3. Note:
   By toggling the keys <Alt> F1, the ladder diagram or equivalent instruction program can be displayed.

Ladder diagram – FLASH1

```
        X0    T1                                              K10
 0     ─┤ ├──┤/├─────────────────────────────────────────────(T0   )─
        T0                                                    K10
 5     ─┤ ├──────────────────────────────────────────────────(T1   )─
        T0
 9     ─┤ ├──────────────────────────────────────────────────(Y0   )─

11     ─────────────────────────────────────────────────────[END  ]─
```

Instruction program – FLASH1

```
 0  LD   X000
 1  ANI  T1
 2  OUT  T0    K10
 5  LD   T0
 6  OUT  T1    K10
 9  LD   T0
10  OUT  Y000
11  END
```

Explanation – FLASH1 instruction program

1. Start of a rung
   (a) Where the first contact on each rung is a normally open contact, then the equivalent instruction will always be:
   LD (Load).

(b) Where the first contact on each rung is a normally closed contact, then the equivalent instruction will always be:
LDI (Load inverse).
2. Contacts in series
Where there are more than one contact connected in series, then to obtain an output, all of the contacts must be correctly operated.
i.e. X0 ON.
   T1 OFF.
Hence, for the timer coil T0 to be energised, input X0 is operated AND the input T1 is not operated. This is written in an instruction program as:
LD   X0.
ANI T0.
Hence, after the first contact on each rung, any additional series-connected contacts will be preceded by the following:
AND for all normally open contacts.
ANI for all normally closed contacts.
3. Outputs
Each rung must be terminated by one or more outputs, i.e.
(a) Output solenoid   Y
(b) Timer coil   T
(c) Counter   C
(d) Internal memory   M
(e) Special instructions, i.e.
    Pulse   PLS
    Master contact   MC
    End of program   END
(f) An advanced instruction
All outputs are preceded with the instruction OUT, followed by the output number and, if required, a constant K value.
i.e. OUT T0.
          K10.
This indicates that timer T0 has been programmed to give an ON time delay of 1 sec.

3.7 Program search

Where a large program has many thousands of steps, then searching for a particular step or determining what devices have been used in the program can become a tedious process.

However, within the Search/Replace function are a number of tools, which make this process much easier.
These are:

1. Find device.
2. Find instruction.
3. Find step number.

4. Find character string.
5. Find contact or coil.
6. Cross-reference list.
7. List of used devices.

Find

The Find option is an extremely useful facility in that it enables:

(a) An immediate jump to a particular step number.
(b) A search for a particular element.

Step numbers

Where a project contains a large number of steps, then it is advantageous to be able to jump to a known part of the program, than have to cursor down from Step 0.
    To use this facility, carry out the following:

1. Let the project FLASH1 be displayed as shown below:

```
        X0      T1                                          K10
0      ─┤ ├────┤/├─────────────────────────────────────────(T0    )─

        T0                                                  K10
5      ─┤ ├──────────────────────────────────────────────── (T1   )─

        T0
9      ─┤ ├──────────────────────────────────────────────── (Y0   )─

11     ─────────────────────────────────────────────────────[END  ]─
```

2. From the main menu, select Find/Replace.
3. Select Find step no.
4. Find step no. window now appears, as shown in Figure 3.6.

Figure 3.6

5. Enter a 5 <OK>.
6. Note that the program immediately jumps to the start of line 5.

7. Hence using this method, any part of the program can be quickly accessed.
8. Repeat the procedure to jump back to the start of the ladder diagram.

Find device

This facility enables a search for an I/O device and Gx-Developer will search for this device and stop at the first match.

1. Let the project FLASH1 be displayed as shown below:

```
       X0    T1                                              K10
  0    ─┤├───┤/├─────────────────────────────────────────────(T0    )─

       T0                                                    K10
  5    ─┤├──────────────────────────────────────────────────(T1    )─

       T0
  9    ─┤├──────────────────────────────────────────────────(Y0    )─

 11    ────────────────────────────────────────────────────[END    ]─
```

2. From the Find/Replace menu, select Find device.
3. The Find device window now appears, as shown in Figure 3.7.

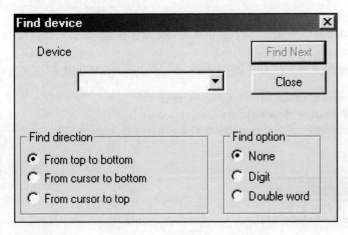

Figure 3.7

4. Enter t0.
5. Select Find Next (Figure 3.8).
6. On the ladder diagram of FLASH1, it can be seen that the coil of T0 is highlighted.

34 *Mitsubishi FX Programmable Logic Controllers*

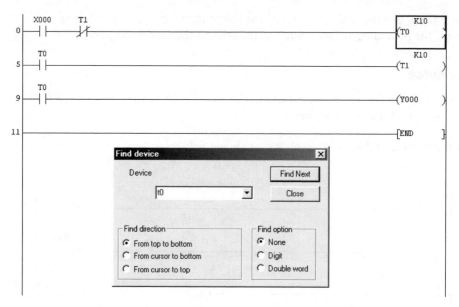

Figure 3.8

7. Selecting Find Next again will cause the next occurrence of T0 to become highlighted, i.e. the normally open contact of T0 at line 5.
8. Select Find Next once more and note the next occurrence of T0 at line 9.
9. Continue selecting Find Next until all of the T0 elements have been found.

Find contact or coil

Instead of searching for both contacts and the coil of a device, it is possible to carry out a search for just either one.

The following describes how a search can be carried out for just the coil of T1:

1. From the Find/Replace menu, select Find contact or coil.
2. The Find contact or coil window now appears.
3. Ensure that Coil has been selected and then enter t1 as shown in Figure 3.9.

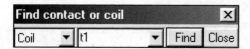

Figure 3.9

4. Select Find.
5. The display now appears as shown in Figure 3.10, with the coil of T1 highlighted.
6. Select Close.

Producing a ladder diagram 35

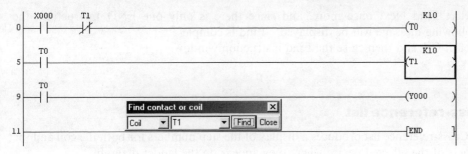

Figure 3.10

Instruction search

Instruction search is an extremely useful facility that enables a search to be carried out, for a particular program instruction.

Hence, where a ladder diagram contains a large number of steps and it is difficult to determine if a particular Instruction is being used, then the instruction search facility can confirm whether or not it is in the program.

The following describes how, using the project FLASH1, a search is carried out for the Instruction END.

1. It will be assumed that the ladder diagram FLASH1 is being displayed.
2. From the main menu, select the following:
   (a) Find/Replace.
   (b) Find instruction.
3. Enter the Instruction end.
4. Select Find Next.
5. The display will now appear as shown in Figure 3.11, with the end instruction highlighted.

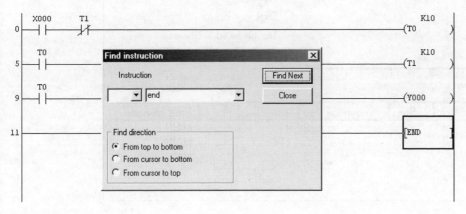

Figure 3.11

6. Select Find Next once more, and since there is only one END Instruction, the following message will be displayed – Find is complete.
7. Select OK and then close the Find instruction window.

Cross-reference list

The Cross-reference list produces a display of the step numbers for both the coil and the contacts of the selected device where they appear on the ladder diagram.

This is very important when fault finding a project and there is a need to track a particular device through the ladder diagram.

The following procedure describes how the Cross-reference details for the Timer t0 in the project FLASH1 are obtained.

1. From the main menu, select Find/Replace.
2. Select Cross-reference list.
3. Enter t0 in the Find device window.
4. Select Execute and all the step numbers of where t0 occurs in the project FLASH1 will be displayed (Figure 3.12).

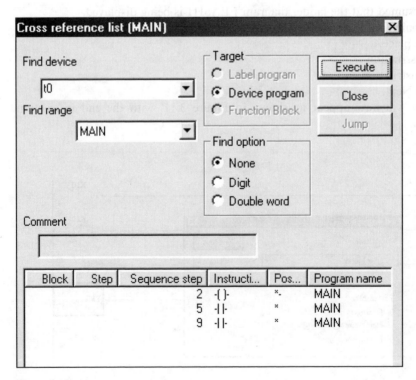

Figure 3.12

Ladder diagram – jump option

From the Cross-reference list, any of the T0 contacts and its coil can be selected and its position on the ladder diagram displayed.

1. On the Cross-reference list, highlight, for example, the T0 contact at Step 5 (Figure 3.13).

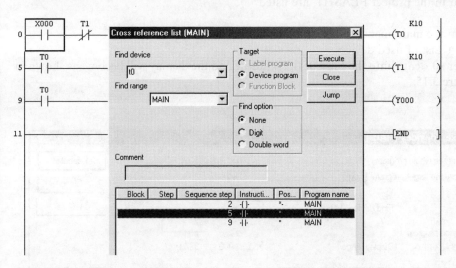

Figure 3.13

2. Select Jump and the cursor will automatically jump to the T0 contact at Step 5 (Figure 3.14).

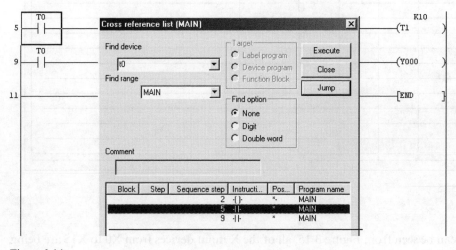

Figure 3.14

3. Select Close to return to the ladder diagram.

List of used devices

Another useful facility, which is in the Find/Replace menu, is the List of used devices. The list enables the user to see what devices are being used in the project.

This is very useful when modifications to the ladder diagram are required, as the programmer will be aware of what devices will be available, for modifying the program.

The following procedure describes how the input X0 and the timers T0 and T1, which are used in the project FLASH1, are listed.

1. From the main menu, select Find/Replace.
2. Select List of used devices.
3. Enter x0 <ent> into the Find device window and the display will become as shown in Figure 3.15.

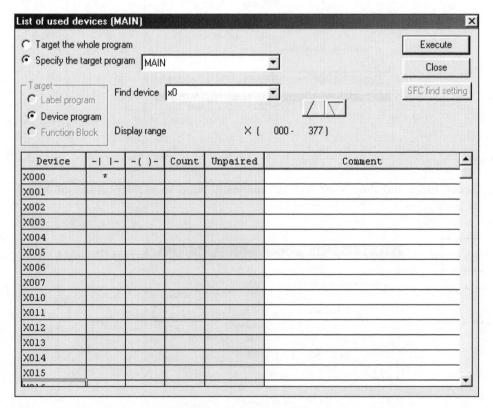

Figure 3.15

4. As can be seen from Figure 3.15, all of the X input devices from X0 to X15 are being displayed.
5. In addition, it can be seen there is a * in the contact column for X0.
6. This indicates that X0 is used in the project FLASH1.

Producing a ladder diagram 39

7. Enter t0 in the Find device window.
8. Select Execute and the display shows that Timers T0 and T1 are being used in the project FLASH1 (Figure 3.16). Hence, the next available timer, which can be used, is T2.

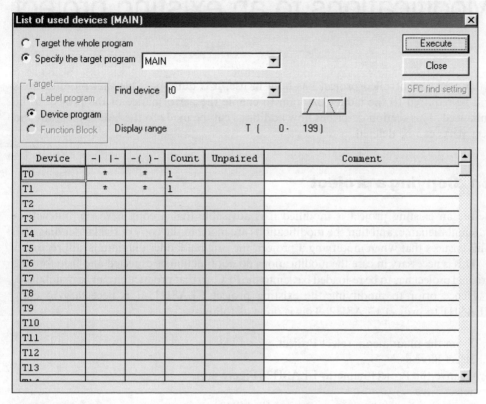

Figure 3.16

4

Modifications to an existing project

After a PLC-controlled project has been in use, it is quite possible that modifications will be required to the ladder diagram to enable the performance of the project to be enhanced. This section describes how additions can be made to the ladder diagram and how parts can be deleted.

4.1 Copying a project

Before an existing project is modified it is advisable that a copy is made, but with a different filename, and that the modifications are made to the copy of the ladder diagram. This ensures that when modifying a project, the original ladder diagram is still retained.

This is necessary in case the modifications do not function as expected and therefore the original project has to be re-loaded back into the PLC, so that production can be maintained.

Hence prior to modifying the existing project FLASH1, it is necessary to copy FLASH1 to project FLASH2. This is done as follows:

1. From the main menu, select Project.
2. Select Save as.
3. Change the Project name to FLASH2 (Figure 4.1).
4. Select Save.
5. Select Yes, to create the new Project FLASH2.
6. The display now appears as shown in Figure 4.2.
7. Note:
   (a) The Project name has changed to FLASH2.
   (b) FLASH1 can still be recalled, whenever required.

Modifications to an existing project 41

Figure 4.1

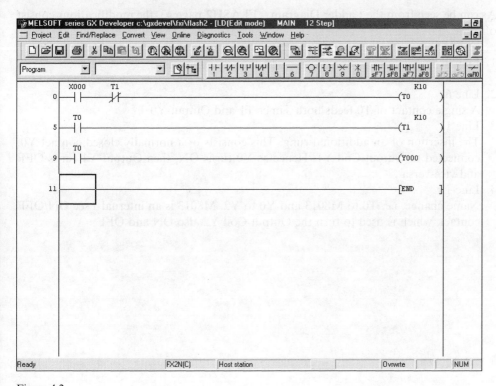

Figure 4.2

4.2 Modification of the ladder diagram FLASH2

Before any modifications can be carried out, it is necessary for the ladder diagram FLASH2 to be displayed on the screen.

At the moment, FLASH2 is identical to FLASH1.

```
       X0      T1                                                    K10
 0    ─┤ ├────┤/├──────────────────────────────────────────────(T0       )─

       T0                                                            K10
 5    ─┤ ├────────────────────────────────────────────────────(T1       )─

       T0
 9    ─┤ ├────────────────────────────────────────────────────(Y0       )─

11    ───────────────────────────────────────────────────────────[END   ]─
```

4.3 Modification details

As can be seen from the Ladder Diagram – FLASH2-page 43, the modifications consist of:

1. Line 0
   The insertion of a normally closed input X1.
2. Line 6
   A single contact of T0 feeds both Timer T1 and Output Y0.
3. Line 11
   The insertion of an additional rung. This consists of a normally closed contact Y0, connected to Output Coil Y1. Hence as Y0 turns ON, then Output Y1 turns OFF and vice-versa.
4. Line 13
   Name change, i.e. T0 to M8013 and Y0 to Y2. M8013 is an internal 1 sec ON/OFF contact, which is used to turn the Output Coil Y2 also ON and OFF.

Modified ladder diagram FLASH2

Insertion of a new contact

1. To insert the normally closed contact X1, between X0 and T1, it will be necessary to change from Overtype mode to INSERT mode.
2. This is done, by pressing the <Insert> key on the keyboard.
3. Note:
   (a) The cursor colour changes to purple.
   (b) The word Insert appears in the bottom right-hand corner of the VDU display (Figure 4.3).
4. Move the cursor to the normally closed contact of T1 on Line 0, by clicking the left-hand mouse button on the T1 contact as shown below:

5. Enter 2 for a normally closed contact.
6. Enter the contact name X1 <ent> (Figure 4.4).

44 *Mitsubishi FX Programmable Logic Controllers*

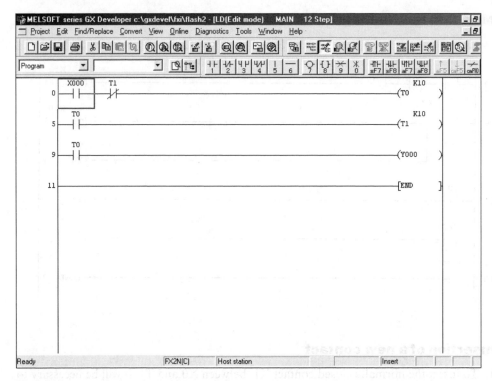

Figure 4.3

Figure 4.4

7. Line 0 will now include the normally closed contact X1.

8. Press F4 to convert the addition of X1.

Partly modified ladder diagram – I

Addition of a branch

To modify the ladder diagram to enable a branch to be added to an existing rung, i.e. to enable the output Y0 to be in parallel with the T0 output, carry out the following:

1. Ensure that the complete ladder diagram is displayed on the screen.
2. Ensure the system is still in Insert mode. If Overtype mode is displayed, then press the <Insert> key to return to Insert mode.
3. Position the cursor on Line 6 and to the right of the T0 contact, as shown in Figure 4.5.

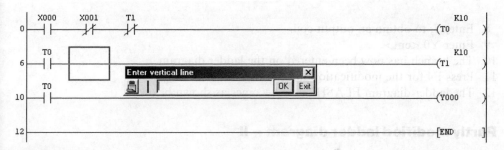

Figure 4.5

4. Enter 5 for a vertical line.
5. Select OK.

6. The display now appears as shown below:

```
     X000    X001    T1                                          K10
0 ───┤ ├────┤/├────┤/├─────────────────────────────────────────(T0    )
     T0    ┌────┐                                                K10
6 ───┤ ├───│    │────────────────────────────────────────────(T1    )
          └────┘
     T0
10 ──┤ ├──────────────────────────────────────────────────────(Y000  )

12 ──────────────────────────────────────────────────────────[END   ]
```

7. Move the cursor down one position.

```
     X000    X001    T1                                          K10
0 ───┤ ├────┤/├────┤/├─────────────────────────────────────────(T0    )
     T0                                                          K10
6 ───┤ ├──────────────────────────────────────────────────────(T1    )
        ┌────┐
        │    │
        └────┘
     T0
10 ──┤ ├──────────────────────────────────────────────────────(Y000  )

12 ──────────────────────────────────────────────────────────[END   ]
```

8. Enter 7, to obtain an output coil.
9. Enter Y0 <ent>.
10. The branch has now been entered on the ladder diagram.
11. Press F4 for the modification to be converted.
12. The ladder diagram FLASH2 now becomes as shown below:

Partly modified ladder diagram – II

```
      X0      X1      T1                                         K10
0 ───┤ ├────┤/├────┤/├─────────────────────────────────────── (T0    )
      T0                                                         K10
6 ───┤ ├──────────────────────────────────────────────────── (T1    )
        │
        └───────────────────────────────────────────────────── (Y0    )

      T0
11 ──┤ ├───────────────────────────────────────────────────── (Y0    )

13 ─────────────────────────────────────────────────────────[END   ]
```

Modifications to an existing project 47

Insertion of a rung

The following describes how a completely new rung can be inserted between Line 6 and Line 11 of the ladder diagram.
  This is done as follows:

1. Ensure the partly modified ladder diagram of FLASH2, as shown on page 46 is being displayed.
2. Place the cursor on Line 11 as shown below:

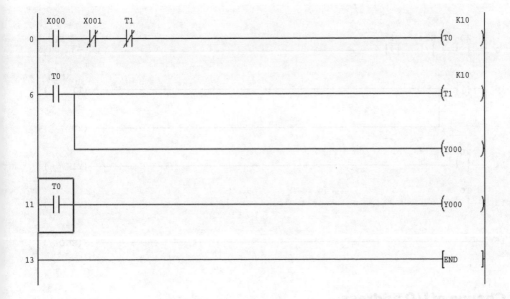

3. From the main menu, select Edit.
4. Select Insert line and the ladder diagram display becomes as shown below:

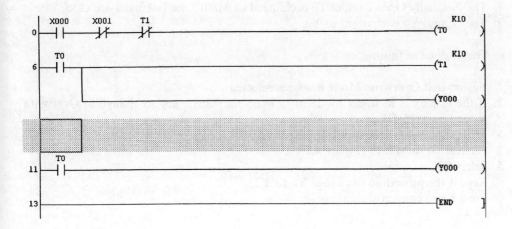

48 *Mitsubishi FX Programmable Logic Controllers*

5. Insert the details for the new line.
   (a) Enter 2 for a normally closed contact.
   (b) Enter the contact name, Y0 <ent>.
   (c) Enter 7 for a coil.
   (d) Enter the coil name, Y1 <ent>.
   (e) Press F4 to convert the modifications.
6. The ladder diagram will now appear as shown below:

Partly modified ladder diagram – III

```
        X0    X1    T1                                           K10
  0    ─┤ ├──┤/├──┤/├──────────────────────────────────────(T0    )─

        T0                                                       K10
  6    ─┤ ├──────────────────────────────────────────────(T1    )─

                                                          ─(Y0    )─

        Y0
 11    ─┤/├──────────────────────────────────────────────(Y1    )─

        T0
 13    ─┤ ├──────────────────────────────────────────────(Y0    )─

 15    ──────────────────────────────────────────────────[END   ]─
```

Change of I/O address

Changing the address of an Input or an Output can be done, by simply overwriting the existing I/O name.

The I/O elements on Line 13 that are to be changed are:

1. The Normally Open contact T0 is changed to M8013, an internal 1-sec clock.
2. The Output Y0 is changed to Y2.

This is done as follows:

1. Ensure that Overwrite Mode has been selected.
2. If the display is in Insert mode, then press the <Ins> key to change to Overwrite mode (see page 43).
3. Move the cursor to the start of Line 13 and double click on the T0 contact (Figure 4.6).
4. Change T0 to M8013 (Figure 4.7).
5. Select OK.
6. Repeat the procedure to change Y0 to Y2.
7. Press F4 to convert the changes.

Modifications to an existing project 49

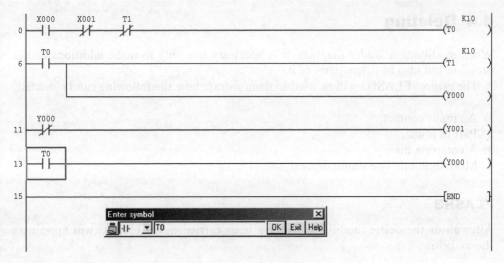

Figure 4.6

Figure 4.7

8. Ladder diagram FLASH2:

   The modified ladder diagram FLASH2, should now be as shown below:

9. Save FLASH2.

4.4 Deleting

When modifying a ladder diagram, it is necessary not only to make additions to the program but also to delete parts of it.

The project FLASH3 will be used to demonstrate how the following can be deleted:

1. An input contact.
2. Part of a line.
3. A complete line.
4. More than one line simultaneously.

FLASH3

After all of the delete modifications have been carried out, FLASH3 will appear as shown below:

```
         X0    T1                                              K10
   0    ─┤ ├──┤/├─────────────────────────────────────────────( T0    )─

   3    ──────────────────────────────────────────────────────[ END   ]─
```

Save FLASH2 to FLASH3

Before carrying out any modifications, save FLASH2 to FLASH3, using the Copy procedure described in Section 4.1.

Deleting an input contact

1. Ensure the project FLASH3 is displayed and it is in Overwrite mode. At this moment in time, FLASH3 will be identical to FLASH2.

```
         X0    X1    T1                                        K10
   0    ─┤ ├──┤/├──┤/├───────────────────────────────────────( T0    )─

         T0                                                    K10
   6    ─┤ ├────────────────────────────────────────────────( T1    )─
              │
              └───────────────────────────────────────────── ( Y0    )─

         Y0
  11    ─┤/├────────────────────────────────────────────────( Y1    )─

         M8013
  13    ─┤ ├────────────────────────────────────────────────( Y2    )─

  15    ──────────────────────────────────────────────────────[ END  ]─
```

Register for eNews, the free email service from Elsevier Science and Technology Books, to receive:

- specially written author articles
- free sample chapters
- advance news of our latest publications
- regular offers
- related event information
- ...and more

Go to **http://books.elsevier.com**, select a subject, register and the eNews will soon be arriving on your desktop!

If you would prefer to register by post, complete and return this card to the address overleaf.

http://books.elsevier.com

ACADEMIC PRESS   MK   Architectural Press   CIMA PUBLISHING   ComputerWeekly PROFESSIONAL SERIES   Digital Press

 Focal Press   G P / P U Gulf Professional Publishing   Newnes   Pergamon Flexible Learning   SYNGRESS

ALL AVAILABLE FROM ELSEVIER

Select the subjects you'd like to receive information about, enter your email and mail address and freepost this card back to us.

ARCHITECTURE AND THE BUILT ENVIRONMENT
- General Architecture
- Architectural Practice Management
- History of Architecture
- Landscape
- Urban design
- Sustainable Architecture
- Planning and Design

BUILDING AND CONSTRUCTION

BUSINESS & MANAGEMENT
- Accounting/CIMA Publishing
- Finance
- Hospitality, Leisure and Tourism
- Human Resources and Training
- Pergamon Flexible Learning
- Knowledge Management
- Management
- Sales and Marketing
- IT Management/Computer Weekly

COMPUTING & COMPUTER SCIENCES
- Computer Weekly:
  - IT Management and Business Computing
- Made Simple Computing:
  - Introduction to computing and programming
- Computer Science:
  - Computer Sciences
  - Artificial Intelligence
  - Computer Graphics
  - Computer Architecture
  - Human Computer Interaction
  - Information Security
  - Information Systems/Databases
  - Networking
  - Operating Systems
  - Software Development
  - Information Systems
- Professional Computing:
  - Data Management
  - Information Security
  - IT Management
  - Networking
  - Operating Systems
  - Software Engineering
  - MCSE certification
  - Security

CONSERVATION AND MUSEOLOGY

ELECTRONICS AND ELECTRICAL ENGINEERING
- Communications
- Control and Instrumentation
- Electrical Power Engineering
- Electronics and Computer Engineering

ENGINEERING
- Bioengineering
- Environmental Engineering
- Industrial Engineering
- Materials Engineering
- Mechanical Engineering
- Optical Engineering
- Petroleum and Petrochemical Processing

FORENSICS

MEDIA TECHNOLOGY
- Film/TV/Video Production
- Postproduction
- Scriptwriting
- Computer Graphics and Animation
- Lighting
- Photography/Imaging
- /Gaming
- Audio
- Radio
- Broadcast and Communication Technology
- Broadcast Management and Theory
- Journalism
- Theatre and Live Performance
- Special effects/Make-up

SECURITY

Name:

Job title:

Email address:

Mail address:

Postcode: **Date:**

Signature:

I would like to receive information by Email ☐   Post ☐   Both ☐

Science and Technology Books, Elsevier Ltd. Registered office: The Boulevard, Langford Lane, Kidlington, Oxon OX5 1GB. Registered number: 1982084

Jo Blackford
Data Co-ordinator
Elsevier
FREEPOST - SCE5435
Oxford
OX2 8BR
UK

FOR CARDS POSTED OUTSIDE UK, PLEASE AFFIX STAMP

As well as conforming to data protection legislation in the way we store your details, Elsevier does not sell or exchange the email/mail address of our subscribers to any other company outside the Reed Elsevier group.

Modifications to an existing project 51

2. On Line 0, move the cursor to the Normally Closed X1 contact.

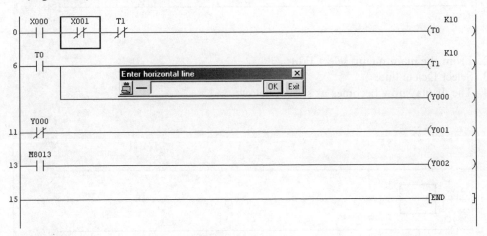

3. Select the horizontal line, i.e. key 6, to overwrite the X1 contact with a horizontal line (Figure 4.8).

Figure 4.8

4. Select OK and the X1 contact will be deleted.
5. Press F4 to convert the modification.

Deleting a branch

The branch at Line 5 will now be deleted.

1. Move the cursor to the branch at Line 5 as shown below:

2. From the main menu, select Edit.
3. Select Delete line.
4. The display now becomes as shown below:

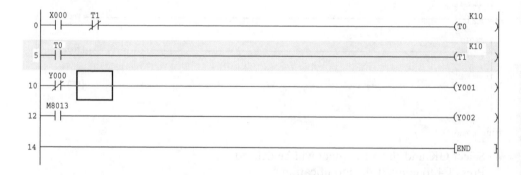

5. Press F4 to convert the modification.
6. Note:
   On earlier versions of Gx-Developer, i.e. GPP Win Version 3, the short vertical line from Line 5 is not automatically deleted. Hence, it is necessary to delete this line, which is done by selecting key 0 (Figure 4.9).

Figure 4.9

Deleting a single line

The single line at Line 5 will now be deleted.

1. Move the cursor to the start of Line 5 as shown below:

```
     X000   T1                                                K10
 0 ──┤ ├───┤/├─────────────────────────────────────────────(T0   )─
     ┌──────┐
     │ T0   │                                                K10
 5 ──┤ ├────┼───────────────────────────────────────────────(T1   )─
     └──────┘
     Y000
 9 ──┤/├───────────────────────────────────────────────────(Y001 )─
     M8013
11 ──┤ ├──────────────────────────────────────────────────(Y002 )─
13 ──────────────────────────────────────────────────────[END   ]─
```

2. Select Edit and then select Delete line.
3. Immediately Line 5 will be deleted.
4. Press F4 to convert the modification and change the Line Numbers.

```
     X0    T1                                                K10
 0 ──┤ ├──┤/├─────────────────────────────────────────────(T0   )─
     Y0
 5 ──┤/├──────────────────────────────────────────────────(Y1   )─
     M8013
 7 ──┤ ├──────────────────────────────────────────────────(Y2   )─
 9 ──────────────────────────────────────────────────────[END   ]─
```

Deleting multiple lines

Use the following procedure where it is required to delete more than one line simultaneously.

The following describes how Lines 5 and 7 are deleted:

1. Place the cursor at the start of Line 5, and holding down the left-hand mouse button, drag the cursor down to the start of Line 7.

2. The display will now appear as shown below, with both the Y0 and M8013 contacts highlighted:

```
     X000    T1                                            K10
0 ───┤ ├────┤/├──────────────────────────────────────(T0      )

     ■Y000■
5 ───■─┤/├─■──────────────────────────────────────────(Y001   )

     ■M8013■
7 ───■─┤ ├─■──────────────────────────────────────────(Y002   )

9 ────────────────────────────────────────────────────[END    ]
```

3. From the main menu, select Edit.
4. Select Delete line and both lines will be deleted simultaneously.
5. Press F4 to convert the modification.
6. Ladder diagram – FLASH3:

The modified Ladder Diagram FLASH3, should now be as shown below:

7. Save FLASH3.

```
     X0      T1                                            K10
0 ───┤ ├────┤/├──────────────────────────────────────(T0      )

5 ────────────────────────────────────────────────────[END    ]
```

5

Serial transfer of programs

5.1 Downloading a project to a PLC unit

The following notes describe how the project FLASH1 is downloaded to an FX2N PLC.

Connection diagram

Connect the computer to the FX2N PLC, as shown in Figure 5.1.

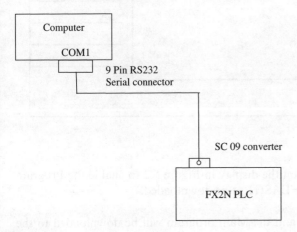

Figure 5.1

Download procedure

1. The SC 09 converter is used to convert the RS232 signals from the computer to the RS 422 format required by the PLC.
2. Ensure the PLC is switched ON and that it is in Stop Mode.
3. Load the project FLASH1 and display the ladder diagram.
4. Connect up the computer to the FX PLC, as shown in the circuit diagram. If a computer is being used, especially a laptop computer, which does not have an RS232 serial connection, refer Section 5.6.
5. From the main menu, select Online.
6. Select Write to PLC.
7. The display now becomes as shown in Figure 5.2.

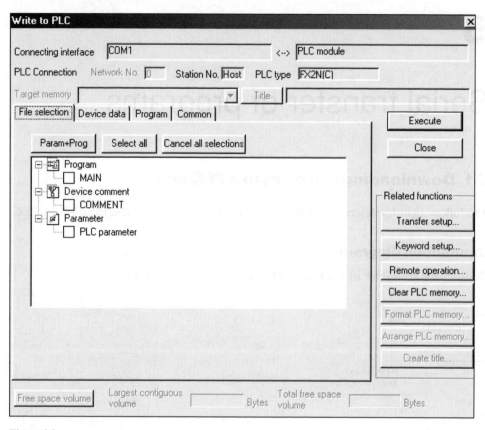

Figure 5.2

8. Select the Param+Prog button on the display in Figure 5.2 to enable the Program and Parameters for the project FLASH1 to be downloaded.
9. Select Execute.
10. Select Yes, and the Parameters and the Main program will be downloaded to the PLC (Figure 5.3).

Write to PLC icon

The Write to PLC icon (Figure 5.4) can be used as an alternative to the drop-down menus.

5.2 Executing the project

To execute the project FLASH1, carry out the following.

1. On the FX2N PLC, switch to RUN.
2. Switch X0 ON.
3. The Y0 Output LED, will now continuously flash ON for 1 sec and then OFF for 1 sec.

Serial transfer of programs

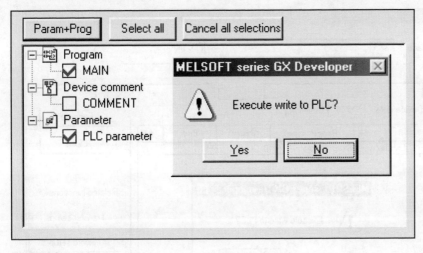

Figure 5.3

Figure 5.4

5.3 Reducing the number of steps transferred to the PLC

When the project FLASH1 was downloaded, the default size of the program that was actually downloaded was 2000 steps. However, as FLASH1 has only 11 steps, then the majority of the information downloaded is 'garbage'.

The number of steps actually downloaded can be reduced from 2000 steps, by using the following procedure:

1. Select Write to PLC.
2. Select the Param+Prog button to enable the Program and Parameters for the project FLASH1 to be downloaded.
3. Select Program.
4. Change the Range type to Step range.
5. Enter 0 for the Start of the step range.
6. Enter 11 for the End of the step range.
7. Note:
   The End step number must be identical to the last step number of the ladder diagram.
8. Select Execute (Figure 5.5).

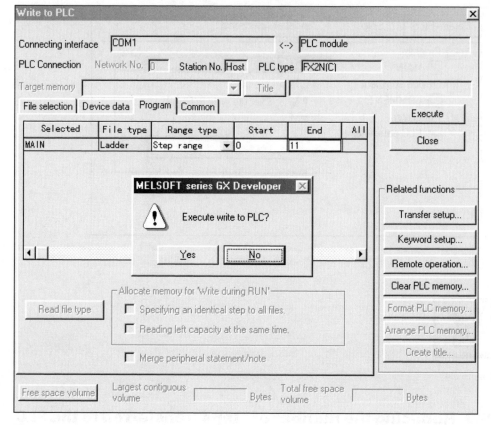

Figure 5.5

9. Select Yes, and the Parameters plus just the 11 steps of FLASH1 will be downloaded to the PLC.
10. Once the download has been completed, select OK followed by Close to return to the ladder diagram.

5.4 Communication setup

If there are any difficulties with downloading the project, then it would be advisable to check the Communication Setup parameters.

1. From the Online Menu, select Transfer setup.
2. The display now becomes as shown in Figure 5.6.
3. Select Connection test.
4. Since the computer is connected to the FX2N PLC, the message 'Successfully connected with the PLC' is displayed.
5. Select OK.

Serial transfer of programs 59

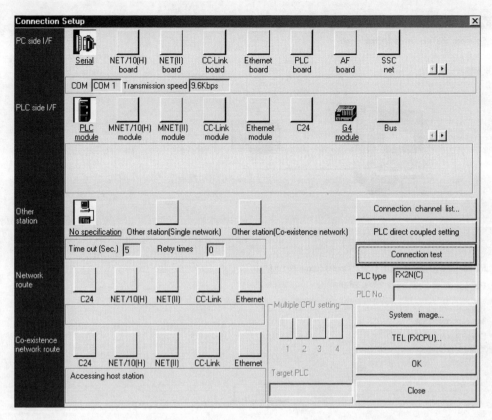

Figure 5.6

5.5 System image

1. Select System image to obtain a pictorial view of the connection setup route.
2. The display now appears as shown in Figure 5.7.
3. As can be seen from the display, the connection setup parameters are:
   (a) Interface   Serial Port using COM1.
   (b) Transmission speed   9.6 kbits per sec.
4. Select OK to close the System image display and return to the connection setup display.

60   *Mitsubishi FX Programmable Logic Controllers*

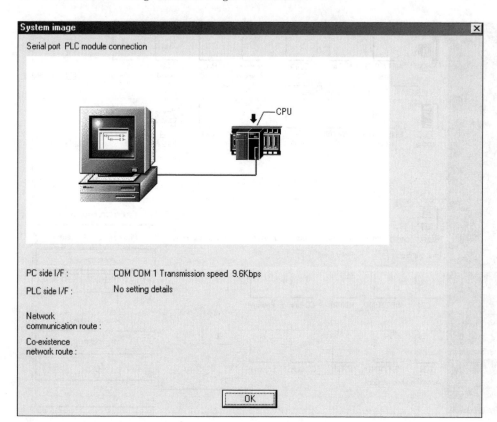

Figure 5.7

5.6 Change of communications port

It may be necessary to change the serial port to another COM port for the following reasons:

1. A serial mouse is using COM1.
2. Some modern laptops do not provide an RS 232 communications port. Therefore, to connect to the SC09 the following alternatives can be used:
   (a) A PCMCIA card with an RS 232 output lead. These cards may only operate via COM2.
   (b) A USB to RS232 adaptor lead. This type of adaptor may only operate via COM4.
3. Note:
   Both the PCMCIA card and the USB adaptor will require the installation of the relevant driver software before they can be used.

Serial transfer of programs 61

If it is necessary to change the communications, i.e. for COM2, then carry out the following:

1. From the Online Menu, select Transfer setup.
2. Double-click on the PC side I/F (Interface), Serial icon.
3. The Serial setting window will now be displayed as shown in Figure 5.8.
4. Click on the COM port down arrow '▼'.
5. Select the required COM port from the list displayed in Figure 5.9.
6. Select OK.
7. Select OK on the communications setup menu and the display will return to the ladder diagram.
8. Select Save to save the COM setting.

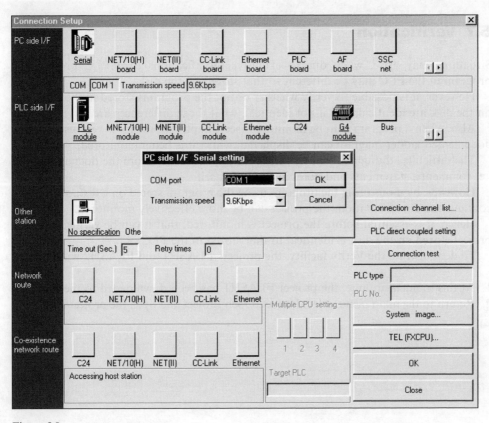

Figure 5.8

62   *Mitsubishi FX Programmable Logic Controllers*

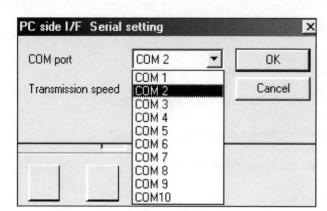

Figure 5.9

5.7 Verification

Situations may arise, when due to extensive modifications to a PLC project, the program in the PLC may be different to that stored on the disk.

However, it is possible to verify whether or not the programs stored in the PLC and on the disk are identical, and if not identical, what those differences are.

Also, when a program is to be monitored (see Chapter 6), then it is very useful if the documented ladder diagram can be displayed, whilst it is being monitored.

The difficulty, though, is that it is not always practical to store the documentation, i.e. comments, statements and notes in the PLC itself.

However, by monitoring the program using the program stored on a disk, which also contains the documentation, the project can be more effectively monitored.

Hence, it is essential before the project is monitored, that it can be verified that the project stored on the disk is identical to that stored in the PLC.

To demonstrate the Verify facility, the projects FLASH1 and FLASH2 will be used.

1. At this moment in time, the project FLASH1 has been downloaded to the PLC.
2. Return to the main menu, select Project, Open Project and open the project FLASH2.

Ladder diagram – FLASH2

```
       X0     X1     T1                                          K10
  0   ─┤├────┤/├────┤/├──────────────────────────────────(T0    )─

       T0                                                        K10
  6   ─┤├──────────────────────────────────────────────────(T1    )─
         │
         └─────────────────────────────────────────────────(Y0    )─

       Y0
 11   ─┤/├──────────────────────────────────────────────────(Y1    )─

      M8013
 13   ─┤├──────────────────────────────────────────────────(Y2    )─

 15   ─────────────────────────────────────────────────────[END   ]─
```

3. Select Online.
4. Select Verify with PLC.
5. Select the Main and Parameter boxes for both the source and the destination. This ensures that the Main program and the Parameters for both FLASH1 and FLASH2 will be verified (Figure 5.10).
6. Select Execute.
7. After the two projects have been verified, any differences will then be displayed (Figure 5.11).
8. As can be seen, the two projects FLASH1 and FLASH2 are quite different. If this situation had occurred with an industrial-based system, it would cause some concern as to why the PLC program and the program saved on disk, were so different.
9. Select Main to return to the ladder diagram FLASH2 (Figure 5.12).

5.8 Uploading a project from a PLC

Circumstances can arise when it is necessary to know what program is stored in the PLC itself. This may be due to a number of modifications being made to the original program and those changes have not been fully documented and saved on the master disks.

Hence, after verifying that the program in the PLC is different to that stored on the disk, the working program within the PLC must be uploaded into Gx-Developer and stored on the master disks.

The following describes how the project FLASH1 is uploaded from the FX2N Series PLC and saved as FLASH4:

1. Select Project as shown in Section 2.1, but with the Project Name FLASH4.
2. Select Online and Read from PLC.

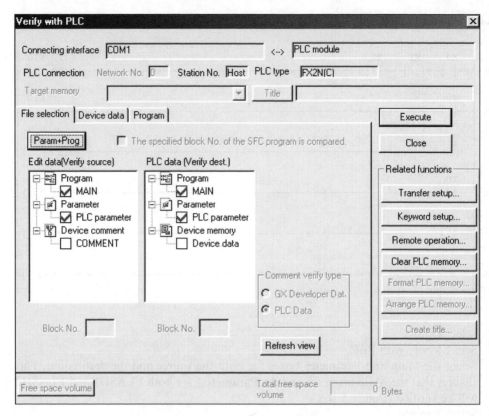

Figure 5.10

Figure 5.11

Serial transfer of programs 65

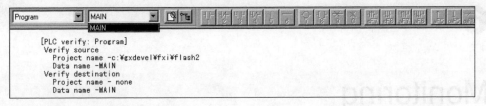

Figure 5.12

3. Alternatively, select the Read from PLC icon, shown in Figure 5.13.

Figure 5.13

4. Select the Param+Prog box (Figure 5.14).

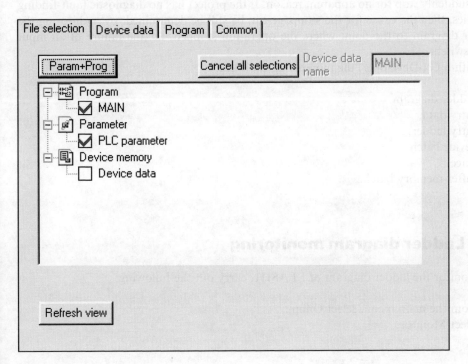

Figure 5.14

5. Select Execute.
6. Select Yes and the PLC program will be uploaded into FLASH4 and displayed.
7. Save FLASH4.

6

Monitoring

Monitoring is an essential tool for determining if a project, new or old, is operating as expected.

For example, during the design and commissioning of a new large-scale PLC-controlled project, it would be necessary to constantly monitor the operation of the project and make changes to the ladder diagram software, should problems arise.

Or, for example, a project may have been working successfully for some years and then suddenly stop for no apparent reason. If the project has no diagnostic fault-finding facilities, then the cause for the stoppage can be still be determined, by monitoring the ladder diagram to the point where the operation stopped, i.e. the failure of an input limit switch.

Within Gx-Developer, the monitoring tools are:

1. Ladder diagram.
2. Entry data.
3. Entry ladder.
4. Device batch.
5. Trace.
6. Buffer memory batch.

6.1 Ladder diagram monitoring

To monitor the ladder diagram of FLASH1, carry out the following:

1. From the main menu, select Online.
2. Select Monitor.

Start monitoring – F3

As can be seen from the display, an alternative to the drop-down menus to start monitoring, is the F3 key.

Start monitoring – icon

A second alternative to start the monitoring process is to select the Monitor mode icon, shown in Figure 6.1.

Figure 6.1

Monitored display – FLASH1

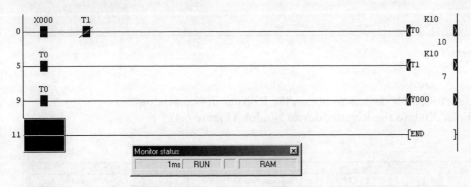

Figure 6.2

Note

1. The display shown in Figure 6.2 shows the ladder diagram FLASH1, whilst in Monitor mode.
2. Beneath the timer outputs, the elapsed times for each timer can be seen.
3. To stop monitoring, press F2 (Write mode).

6.2 Entry data monitoring

Entry data monitoring is an alternative method for monitoring the conditions of the ladder diagram elements. It enables far more information to be obtained, than that displayed on the ladder diagram.

To monitor using entry data monitoring, carry out the following:

1. From the main menu, select Online.
2. Select Monitor.
3. Select Entry Data Monitor.
4. The display now becomes as shown in Figure 6.3.

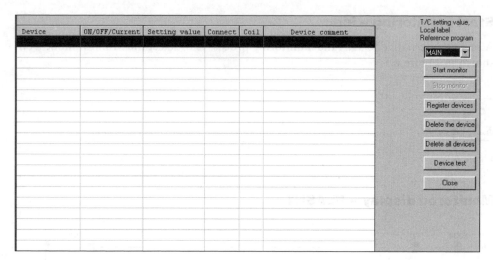

Figure 6.3

5. Select Register devices to obtain the Register device window.
6. Enter X0 into the Register device window (Figure 6.4).

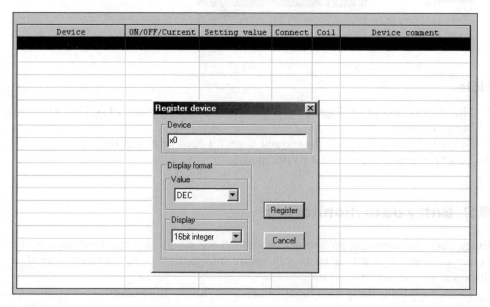

Figure 6.4

7. Press <ent>.
8. The entered element, i.e. X0 will now be transferred to the Monitor window (Figure 6.5).

Monitoring 69

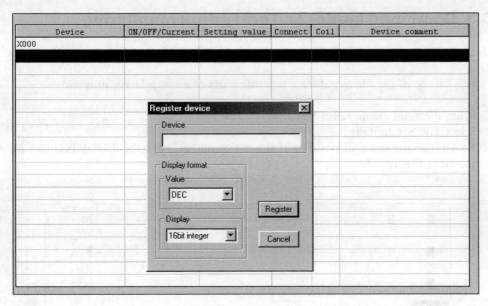

Figure 6.5

9. Enter the additional following elements:
   (a) T0.
   (b) T1.
   (c) Y0.
10. Select Cancel.
11. Ensure that PLC is switched to RUN.
12. Select Start monitor and the selected elements of FLASH1 will now be monitored in Data Entry Monitor mode (Figure 6.6).

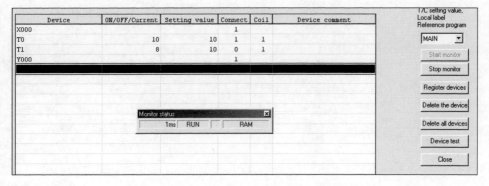

Figure 6.6

70 *Mitsubishi FX Programmable Logic Controllers*

6.3 Combined ladder and entry data monitoring

Using Windows, it is possible to monitor both the ladder diagram and the entry data.

This can be of use for monitoring separate parts of a much larger program.

1. From the main menu, select Window.
2. Select Tile horizontally.
3. Re-arrange the size and position of the two windows, until the display becomes as shown in Figure 6.7.

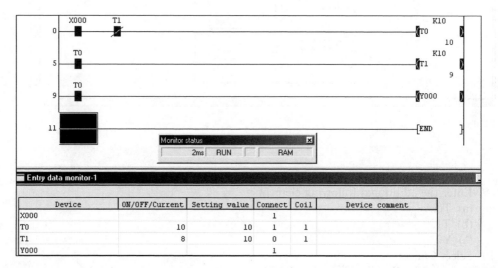

Figure 6.7

4. Note the ladder diagram and the entry data are both being monitored simultaneously.
5. Close the Entry data monitor window.

7

Basic PLC programs

To gain experience of using the Gx-Developer software and the FX PLCs, a range of fairly basic programs will now be developed and tested.

These PLC programs could be used to replace conventional systems consisting of relays timers and counters.

The following is a list of such programs, which will be used in this chapter:

1. TRAF1   – Traffic light controller.
2. FURN1   – Furnace temperature controller.
3. INTLK1 – Safety interlock circuit.
4. LATCH1 – Mains failure latch circuit.
5. COUNT1 – Extended time delay.
6. COUNT2 – Online programming.
7. BATCH1 – Batch counter.
8. BATCH2 – Assignment.
9. MC1   – Master control.

7.1 Traffic light controller – TRAF1

Task

Using Gx-Developer, produce a PLC ladder diagram, which can enable an FX2N PLC to simulate a simple combinational logic traffic light controller.

Block diagram

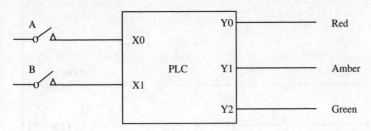

Figure 7.1

Determined by the condition of the input switches A and B, the output lights will turn ON/OFF according to the following truth table.

Note

If the Input switch or the Output is OFF, = logic 0
If the Input switch or the Output is ON, = logic 1

Truth table

Inputs		Outputs		
A	B	Red	Amber	Green
(X0)	(X1)	(Y0)	(Y1)	(Y2)
0	0	1	0	0
0	1	1	1	0
1	0	0	0	1
1	1	0	1	0

Solution

From the truth table, the following ladder diagram can be produced:

```
       X0    X1                                         RED
  0  ──┤/├──┤/├─────────────────────────────────────────(Y0 )─
       X0    X1
     ──┤/├──┤ ├─
       
       X0    X1                                         AMBER
  6  ──┤/├──┤ ├─────────────────────────────────────────(Y1 )─
       X0    X1
     ──┤ ├──┤ ├─

       X0    X1                                         GREEN
 12  ──┤ ├──┤/├─────────────────────────────────────────(Y2 )─

 15  ──────────────────────────────────────────────────[END ]─
```

By inspection it can be seen that the ladder diagram can be reduced to the one shown below:

```
       X0    X1                                           RED
0     ─┤/├──┤/├──────────────────────────────────────────(Y0  )─
             │
             X1
            ─┤ ├─

       X0    X1                                           AMBER
6     ─┤/├──┤ ├──────────────────────────────────────────(Y1  )─
       │
       X0
      ─┤ ├─

       X0    X1                                           GREEN
12    ─┤ ├──┤/├──────────────────────────────────────────(Y2  )─

15    ──────────────────────────────────────────────────[END ]─
```

Final modification

The final modification is based on the fact that where there are two contacts, one normally open and one normally closed connected in parallel and having the same name, i.e. X0, then the two contacts can be replaced with a single link.

```
    X0
   ─┤ ├─
    │
    X0             is identical to    o───────────o
   ─┤/├─
```

Ladder diagram – TRAF1

The final circuit for the ladder diagram TRAF1 is shown below:

```
       X0    A                                            RED
0     ─┤/├──┤ ├──────────────────────────────────────────(Y0  )─

       X1    B                                            AMBER
2     ─┤ ├──┤ ├──────────────────────────────────────────(Y1  )─

       X0    A    X1    B                                 GREEN
4     ─┤ ├──┤ ├──┤/├──┤ ├─────────────────────────────────(Y2  )─

7     ──────────────────────────────────────────────────[END ]─
```

Hence, it can now be seen that

Red = $\overline{A}$.
Amber = B.
Green = $A.\overline{B}$.

The above can also be proved theoretically, using either Boolean Algebra or by using Karnaugh Mapping techniques.

Principle of operation

1. Line 0
   When there is no input on X0, i.e. $A = 0$, then the Red output Y0, will be ON. Conversely, when $A = 1$, then the Red output will be OFF.
2. Line 2
   The B input X1 is directly connected to the Amber output Y1, hence producing the truth table below:

B input (X1)	Amber output (Y1)
0	OFF
1	ON

3. To turn the Green output Y2 ON, then input A should be ON whilst input B should remain OFF.

7.2 Furnace temperature controller – FURN1

A furnace has to be controlled from cold between the limits of:

1. A low temperature setting (tl).
2. An upper temperature setting (tu).

This technique ensures that the furnace will not be turned ON/OFF as often as it would be, if only a single thermostat had been used.

This ensures that the life of the electric elements within an electric furnace will be extended.

Similarly, if it were a gas-fired furnace, then the life of the gas burners would also be extended.

Block diagram

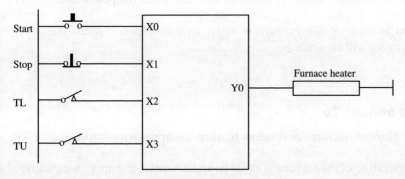

Figure 7.2

While heating

Let t be the temperature of the furnace.

    t < tl   furnace ON.
    t < tu   furnace ON.
    t >= tu   furnace OFF.

While cooling

    t < tu   furnace OFF.
    t <= tl   furnace ON.

Temperature response characteristic

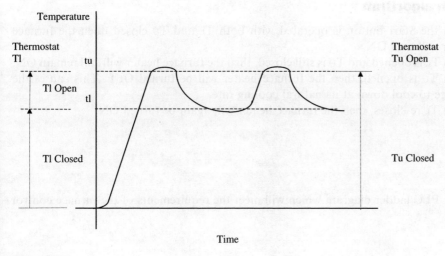

Figure 7.3

Temperature sensor T_L

When the temperature of the furnace is below tl, then the lower temperature sensor T_L will be closed.

When the temperature of the furnace is equal to and above tl, then the lower temperature sensor T_L will be open.

Temperature sensor T_U

When the temperature of the furnace is below tu, then the upper temperature sensor T_U will be closed.

When the temperature of the furnace is equal to and above tu, the upper temperature sensor T_U will be open.

Safety procedures

1. The type of thermostats used, have their contacts closed, when the temperature is below their set value. This ensures that if the connections to the thermostats or the thermostats themselves, become open circuit, the furnace will switch OFF, i.e. a FAIL SAFE condition.
2. Also for safety purposes, the STOP button X1, must be normally closed. This ensures if the wire connecting to the STOP button breaks, the system will again FAIL SAFE. Hence, on the ladder diagram, the STOP input is entered as a normally open contact.
3. For further details on safety, refer Chapter 10.

System algorithm

1. When the Start button is operated with both T_L and T_U closed, then the furnace heater will turn ON.
2. When T_L is opened and T_U is still closed, then the furnace heater will still remain ON.
3. When T_U is opened, then the furnace heater will be turned OFF. This causes the furnace to cool down at its natural cooling rate.
4. When T_L re-closes, then the furnace heater will again be turned ON.

Task

Design a PLC ladder diagram which will meet the requirements of the furnace control system.

Ladder diagram – FURN1

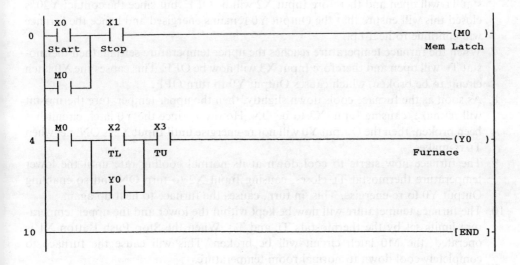

Note

To enter M0, after all of Line 0 has been entered, carry out the following:

1. Place the write cursor below X0.
2. Enter 3 for a single normally open parallel contact.
3. Enter M0 <ent>.
4. Repeat the same procedure at Line 4, i.e. complete the top rung and then under X2, enter:
   (a) 3   for a single parallel contact.
   (b) Y0 <ent>   for the normally open contact Y0.

Principle of operation

1. Line 0
   The normally open contact X0 and the normally open contact X1, which is connected to the normally closed contacts of the STOP BUTTON and the internal Memory Coil M0, form a Start/Stop latch circuit.
2. The operation of the Start Push Button X0 will cause M0 to energise and latch via its own contact.
3. Line 4
   When the Furnace is cold, then the thermostats T$_L$ and T$_U$ will be closed and therefore Inputs X2 and X3 will be ON.
4. With the operation of X0 and hence the operation of M0, the circuit is complete and hence the Output Y0 will turn ON.

5. With the operation of Y0, the furnace will now start to heat up.
6. When the furnace temperature reaches the lower temperature setting, then thermostat TL will open and therefore Input X2 will be OFF, but since the contact Y20 is closed this will ensure that the Output Y0 remains energised and hence the furnace will continue to heat up.
7. When the furnace temperature reaches the upper temperature setting, then thermostat TU will open and therefore Input X3 will now be OFF. This causes the Y0 latch circuit to be broken, which causes Output Y0 to turn OFF.
8. As soon as the furnace cools down slightly, then the upper temperature thermostat will re-make, causing Input X3 to be ON. However, since the Y0 latch circuit has been broken, then the Output Y0 will not re-energise until Input X2 is ON, i.e. when TL re-makes.
9. The furnace now starts to cool down at its normal cooling rate until the lower temperature thermostat TL closes, causing Input X2 to turn ON and so enabling Output Y0 to re-energise. This, in turn, causes the furnace to heat up again.
10. The furnace temperature will now be kept within the lower and the upper temperature limits set by the thermostats TL and TU. When the Stop Push Button X1 is operated, the M0 latch circuit will be broken. This will cause the furnace to completely cool down to normal room temperature.

7.3 Interlock circuit – INTLK1

Description

An automatic sauce blending control system, which consists of three main sections, has to be operated manually.

S.No.	Section	Input (Manual)	Output
1.	Hopper input	X0	Y0
2.	Weigher output	X1	Y1
3.	Blender output	X2	Y2

As part of the manual operation of the system, the Inputs X0, X1 and X2 are connected to push buttons, to enable the Outputs Y0, Y1 and Y2 to be manually operated.

However, it is an essential requirement of the design that an interlock circuit be included, which will ensure that only one of the Outputs can be ON at any one time.

Blending system diagram

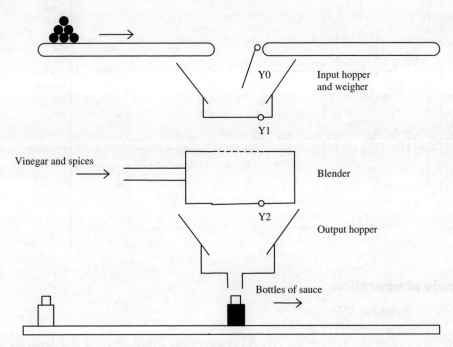

Figure 7.4

Task

Produce a PLC ladder diagram which ensures that only one of the Output Solenoids can be ON, at one time.

X0 is a normally open push button, which is used to energise and latch the Hopper Input Solenoid Y0.

X1 is a normally open push button, which is used to energise and latch the Blender Input Solenoid Y1.

X2 is a normally open push button, which is used to energise and latch the Blender Output Solenoid Y2.

X3 is a normally closed push button/switch, which is used to Reset any latched Output.

Ladder diagram – INTLK1

```
         X0    X1    X2   Reset
                           X3
 0      ─┤├──┤/├──┤/├──┤├──────────────────(Y0)─  Hopper
         │                                         Input
         Y0
        ─┤├─

         X1    X0    X2    X3
 6      ─┤├──┤/├──┤/├──┤├──────────────────(Y1)─  Blender
         │                                         Input
         Y1
        ─┤├─

         X2    X0    X1    X3
12      ─┤├──┤/├──┤/├──┤├──────────────────(Y2)─  Blender
         │                                         Output
         Y2
        ─┤├─

18      ──────────────────────────────────[END]─
```

Principle of operation

1. Switch the Reset X3, ON.
2. The basic circuit consists of three latch circuits.
3. When any one of the Inputs X0, X1, X2 is operated, it will energise and latch its corresponding Output Solenoid.
4. At the same time, if there is another output which has been previously energised and latched, then the normally closed contact of the input, which has just operated, will break the latch to that particular output. Hence, only one output can be on, at any one time.
5. The operating of X3 will enable the last output to have been latched to now be de-energised.

7.4 Latch relays

Latch relays are used to ensure that should there be an interruption in the voltage supply, either due to a mains failure or a fault in the DC power supply, it will still be possible for the program to continue execution from the same point it was at, when the interruption occurred.

The latch relays use battery backup, to retain their ON/OFF condition, whenever there is an interruption to the voltage supply.

Latch memory range

The Latch memory range on the FX2N PLC is M500–M1535.

Basic PLC programs 81

Ladder diagram – LATCH1

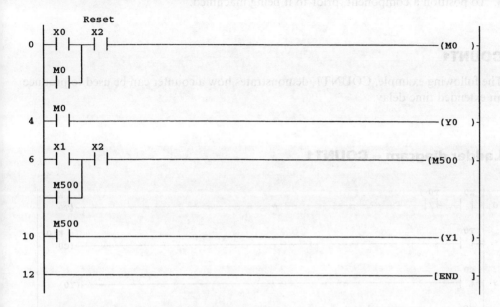

Principle of operation – LATCH1

1. Enter, test and save the project Latch1.
2. Download the project to the FX2N PLC.
3. Ensure the project is working correctly using the inputs X0, X1 and X2.
4. Line 0
   Input X0 operates M0, which then latches via its own contact and also energises Output Y0.
5. Line 6
   Input X1 operates M500, which then latches via its own contact and also energises Output Y1.
6. Reset
   Opening Input X2 resets both latch circuits, provided Inputs X0 and X1 are OFF.
7. With Input X2 ON, momentarily operate Inputs X0 and X1, so that both Outputs Y0 and Y1 are turned ON.
8. Momentarily turn OFF the 240 V mains supply to the FX2N PLC.
9. When the mains supply is turned back ON, then only the Output Y1 will be ON. This is because M500 remained latched ON, due to it having battery backup.

7.5 Counters

Counters are a very important part of a sequence control system.
They can be used for example:

1. To ensure a particular part of a sequence is repeated a known number of times.
2. To count the number of items being loaded into a carton.

82 Mitsubishi FX Programmable Logic Controllers

3. To count the number of items passing along a conveyor belt, in a given time.
4. To position a component, prior to it being machined.

COUNT1

The following example, COUNT1, demonstrates how a counter can be used to produce an extended time delay.

Ladder diagram – COUNT1

```
        X0      T0                                              K15
  0    ─┤ ├────┤/├─────────────────────────────────────────────(T0  )─

        T0                                                      K10
  5    ─┤ ├──────────────────────────────────────────────────── (C0  )─

        C0
  9    ─┤ ├────────────────────────────────────────────────────(Y0  )─

        X0
 11    ─┤ ├──────────────────────────────────────────[PLS   M0   ]─

        M0
 14    ─┤ ├──────────────────────────────────────────[RST   C0   ]─

 17    ─────────────────────────────────────────────────[END ]─
```

Note

1. To enter -[PLS M0]-, enter the following:
   (a) pls   <space>.
   (b) m0   <ent>.
2. Use the same procedure for -[RST C0]-, i.e:
   (a) rst   <space>.
   (b) c0   <ent>.

Principle of operation

1. Line 0
   The closing of Input X0, and the normally closed timer contact T0, will provide a path to enable the coil of Timer T0 to be energised. After 1.5 sec, Timer T0 times out and its normally closed contact will open, causing the timer to become de-energised for a time equal to one scan period, which for COUNT1 will be approximately

1.5 msec. With the timer dropping-out, its contact re-closes causing the timer to be re-energised once more. This 'cut-throat' timer circuit is effectively a pulse oscillator, whose contacts momentarily operate every 1.5 sec.

2. Line 5

With the momentary closure of the normally open contacts of T0, a count pulse is sent to Counter C0 every 1.5 sec (Figure 7.5).

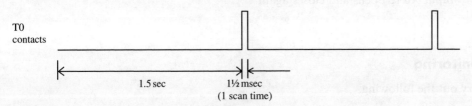

Figure 7.5

3. Line 9

Counter C0 counts the incoming pulses, and when the number of pulses equals the preset K value, i.e. 10, all the C0 contacts operate as follows:
(a) All normally open contacts, CLOSE.
(b) All normally closed contacts, OPEN.
The normally open contact C0 closes, hence energising the Output Coil Y0. Therefore, the circuit gives an output signal on Y0, 15 sec after the Input X0 closes. Hence, the circuit can be considered as an extended timer.

4. Line 11

Whenever Input X0 closes, this energises a special output, which is known as a pulse circuit, PLS. A pulse circuit only operates on the closing of an input, and when energised, the pulse circuit will cause its associated output the Internal Memory M0, to energise for a time equal to 1 scan time for the program. Hence, the contacts of M0 will be closed, for approximately 1.5 msec.

5. Line 14

(a) PLS waveforms

The waveforms associated with the PLS circuit are shown in Figure 7.6.

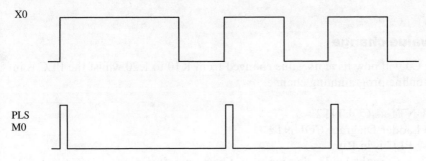

Figure 7.6

(b) From the above waveforms, it can be seen that each time Input X0 operates, the Instruction PLS M0 will be executed and the normally open contact of M0 will momentarily close, hence causing the Counter C0 to be reset to zero.
(c) Hence, with the operation of Input X0 and the resetting of Counter C0, the cycle will repeat itself.
(d) Even though Input X0 remains closed, the pulse circuit will not re-operate until Input X0 re-opens and closes again.

Monitoring

Carry out the following:

1. Open a new project and give it the name, COUNT1.
2. Enter the ladder diagram.
3. Save the program.
4. Download to the FX2N PLC.
5. Monitor the Ladder Diagram COUNT1.
6. Press F3 to start monitoring.

7.6 Online programming

Using the online programming facility of Gx-Developer, it is possible to modify one block at a time of the project, even though the PLC is in RUN.

In a continuous process, which cannot be stopped, i.e. in a steel works, online programming may be the only way that changes to the program can be carried out.

However, online programming can be dangerous, since once the modifications have been entered, they become operative on the next scan of the program.

The project COUNT2, is used to demonstrate the use of the online programming facility.

Counter value change

The counter C0 will now have its value changed from K10 to K20 whilst the PLC is in Run, i.e. an online programming change.

1. Save COUNT1 as COUNT2.
2. Open the Ladder Diagram COUNT2.
3. Ensure the PLC is in Run.
4. Press F2 to ensure the ladder diagram is not being monitored.

Basic PLC programs 85

5. Move the cursor to Line 5 and over the output -(C0^{K10})- as shown below:

```
     X000    T0                                                     K15
0   ─┤├─────┤/├─────────────────────────────────────────────────────(T0    )

     T0                                                             K10
5   ─┤├─────────────────────────────────────────────────────────────(C0    )

     C0
9   ─┤├─────────────────────────────────────────────────────────────(Y000  )

     X000
11  ─┤├─────────────────────────────────────────────────[PLS    M0        ]

     M0
14  ─┤├─────────────────────────────────────────────────[RST    C0        ]

17  ─────────────────────────────────────────────────────────[END         ]
```

6. Double-click the left-hand mouse button to obtain the output information (Figure 7.7).

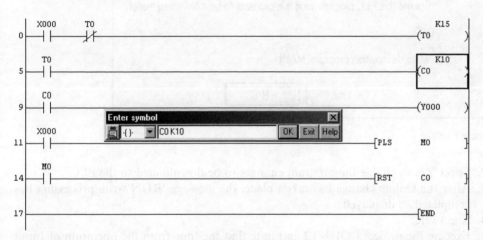

Figure 7.7

7. Change the output details as in Figure 7.8.

Figure 7.8

8. Select OK and the display will appear with all of Line 5 greyed out.
9. From the main menu, select Convert (Figure 7.9).

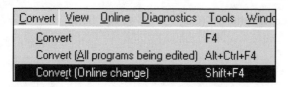

Figure 7.9

10. Select Convert (Online change).
11. The message shown in Figure 7.10 is now displayed.

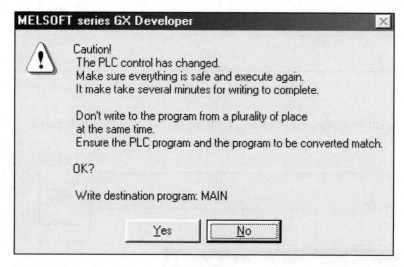

Figure 7.10

12. Select Yes to enable the program changes to be downloaded to the PLC.
13. After the Online change has taken place, the message 'RUN write processing has completed' is displayed.
14. Select OK.
15. Execute the project COUNT2 and note that the time from the operation of Input X0 to when Y0 turns ON, will now be 20 sec.
16. Save COUNT2.

7.7 Batch counter – BATCH1

A batch counter is a counter, which turns a specified output ON after counting a required number of input pulses.

The specification for BATCH1 is as follows:

1. Use Input X0, to reset Counter C0.
2. Use Input X1, to input pulses to the counter.

3. After 10 input pulses, Output Y0 is to come ON.
4. Resetting the counter will turn the Output Y0 OFF and enable counting to be repeated.

Batch counter – BATCH1

```
        X0
0  ─────┤ ├──────────────────────────────────────────[RST   C0    ]─
                                                              K10
        X1
3  ─────┤ ├──────────────────────────────────────────────(C0      )─
        C0
7  ─────┤ ├──────────────────────────────────────────────(Y0      )─

9  ──────────────────────────────────────────────────────[END     ]─
```

Principle of operation

1. Line 0
   The momentary operation of Input X0 will Reset the Counter C0. This, in turn, will cause Output Y0 to turn OFF (Line 7).
2. Line 3
   Each time Input X1 closes, this will increment the contents of C0.
3. Line 7
   After Input X1 has been pulsed 10 times, the C0 contacts will operate, to enable the Output Y0 to energise.

7.8 Assignment – BATCH2

Modify BATCH1 so that the following sequence can be obtained:

1. After the count value of 10 has been reached, then Output Y0 will be energised and stay ON for just 5 sec.
2. After the 5-sec delay, the following is to occur:
   (a) Output Y0 will turn OFF.
   (b) Counter C0 will automatically be reset.
3. Counter C0 can then count up once more to 10 and Output Y0 will again be energised and stay ON, for 5 sec.

7.9 Master control – MC1

The master control function enables sections of a ladder diagram to be enabled/disabled.

In the example below, the section consisting of Lines 9–23 cannot be executed, until the master control Instruction -[MC N0 M0]- is executed.

The serial master contacts N0 ╪ M0 cannot be programmed in directly.

They will appear automatically, when the ladder diagram is monitored for the first time.

Ladder diagram – MC1

```
          Start Stop Air  Oil  Guard
          X0    X1   X2   X3   X4
    0 ────┤├────┤/├──┤├───┤├───┤├──────────────────[MC    N0    M0 ]─
          │
          M0
          ├┤
          
  N0 ─────┬── M0
          │
          M8013  1 second pulses
    9 ────┤├──────────────────────────────────────────────(Y0  )─
          
          Y0
   11 ────┤/├────────────────────────────────────[ALTP  Y1  ]─
          
          Y1
   15 ────┤/├────────────────────────────────────[ALTP  Y2  ]─
          
   19 ─────────────────────────────────────────────[MCR   N0  ]─
          
          X7
   21 ────┤├──────────────────────────────────────(M8034  )─
          
   24 ─────────────────────────────────────────────────[END ]─
```

Principle of operation

1. In this example, the master control is controlling an LED output display, which indicates which section of a machine tool cycle is currently being executed.
2. Since the X Inputs to the Instruction -[MC N0 M0]- are controlling this display section, they can be regarded as Primary Inputs, i.e. they are controlling more than just a single output.
3. The primary inputs have been given the following names:
   (a) X0   Start.
   (b) X1   Stop.
   (c) X2   Air.
   (d) X3   Oil.
   (e) X4   Guard.

4. Line 0
   (a) When all of the primary circuits X1–X4 are made, then the momentary operation of the Start button X0 will enable the Instruction -[MC N0 M0]- to be executed.
   (b) The M0 contact, which is in parallel with X0, will close and this will provide a latch circuit to the master control instruction, when the Start button is released.
   (c) However, the main function of the MC instruction is to control a section of the ladder diagram. The control operation is shown diagrammatically, as a pair of normally open contacts N0 ╪ M0, on the left-hand side of the ladder diagram.
   (d) These contacts effectively control the operation of the ladder diagram from Lines 9–19.
   (e) At Line 19, the Master Control Reset Instruction -[MCR N0]- terminates the operation of the master control for this section of the ladder diagram.
5. The program uses special M Coil M8013 to simulate the signals, which would indicate which section is currently being executed. This special M Coil (Figure 7.11) provides an accurate 1-sec clock signal, for any ladder diagram project.

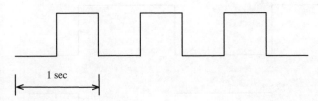

Figure 7.11

6. Line 9
   With the operation of the master control, the 1-sec M8013 contacts will continuously pulse Y0 ON for 0.5 sec and then OFF for 0.5 sec.
7. Line 11
   The Instruction, -[ALTP Y1]-, causes Y1 to alternatively turn ON and then OFF, each time the inverted contacts of Y0 close. Hence, Y1 turns ON for 1 sec and then OFF for 1 sec.
8. Line 15
   The inverted 1-sec ON/OFF pulses from Y1 will now enable the Instruction -[ALTP Y2]- to continuously turn the Output Y2 ON for 2 sec and then turn it OFF for 2 sec.
9. Line 19
   The Instruction -[MCR N0]- terminates the master control section. This ensures that the remaining section of the ladder diagram operates irrespective of whether the master control contacts are open or closed.
10. Line 21
    (a) The Input X7 is used to simulate a fault condition.
    (b) The operation of X7 will energise the special M Coil M8034 irrespective of the operation of the master control.
    (c) The function of M8034 is that, on being operated, it will turn all of the outputs OFF.

11. When the program is executed, it will be seen that with X7 OFF, the output LED display of Y0, Y1 and Y2 is a continuous incrementing binary pattern from 000 to 111, i.e. from 0 to 7.

Waveforms

Figure 7.12

Note

At Output No. 6, Y2 = 1, Y1 = 1, Y0 = 0.

Now, $110_2 = 6$, which confirms that the Outputs Y2, Y1 and Y0 are outputting an incrementing binary pattern.

8

PLC sequence controller

The following notes describe how a system, which consists of two pneumatic pistons, can be controlled using a Mitsubishi FX PLC.

The PLC is required to operate two single-acting electrically actuated pneumatic pilot valves, which in turn control the two pneumatic pistons.

Basic system

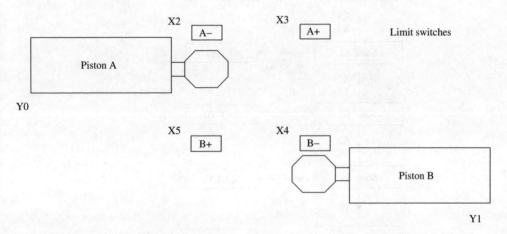

Figure 8.1

Sequence of operation

The sequence of operations for the two pistons is as follows:

1. Start of sequence.
2. A+   Piston A OUT.
3. B+   Piston B OUT.
4. A−   Piston A IN.
5. 5-sec time delay
6. B−   Piston B IN.
7. End of sequence

92 Mitsubishi FX Programmable Logic Controllers

8.1 Sequence function chart – SFC

A sequence function chart is a pictorial representation of the system's individual operations, which when combined show the complete sequence of events.

Once this diagram has been produced, then from it, the corresponding ladder diagram can be more easily designed.

The Gx-Developer software, which has as the facility for programming directly in SFC, is not described in this book.

Sequence function chart – PNEU1

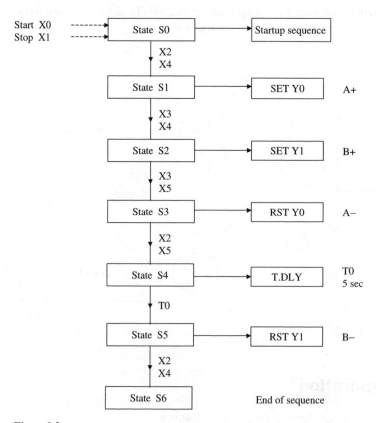

Figure 8.2

Description – sequence function chart

1. The sequence function chart consists, basically, of a number of separate sequentially connected states, which are the individual constituents of the complete machine cycle, that controls the system. An analogy is that each state is like a piece of a jigsaw puzzle; on its own it does not show very much, but when all the

pieces are correctly assembled, then the complete picture is revealed. Each state has the following:
(a) An input condition.
(b) An output condition.
(c) A transfer condition.

When the input condition into a state is correct, then that state will produce an output condition.

That is, an output device or devices will be:
(a) Turned ON and remain ON.
(b) Turned OFF and remain OFF.
2. When the output or outputs are turned ON/OFF, then the system's input conditions will change to produce a transfer condition.
3. The transfer condition is now connected to the input condition of the next sequential state.
4. If the new input condition is correct, then the sequence moves to the next state.
5. From the sequence function chart for PNEU1, it can be seen that when the start push button is operated, this is the input condition for state 0.
6. The output condition from State 0 is the startup sequence, which will reset both Solenoid A and Solenoid B. With Inputs X2 and X4 now made, the transfer from State 0 can take place.
7. The transfer conditions from State 0 are the correct input conditions for State 1, and hence the process now moves from State 0 to State 1.
8. The process will now continue from one state to the next, until the complete machine cycle is complete.
9. From the sequence function chart, the ladder diagram can now be produced.

8.2 Ladder diagram – PNEU1

It will now be found that the production of the PLC ladder diagram, from the sequence function chart, will be a far easier process.

Ladder diagram – PNEU1

```
   X0    X1    T0   (State S5)
   ─┤├──┬─┤├──┤/├──────────────────────────[MC    N0    M0   ]─  Start
   M0   │                                                         Up
   ─┤├──┘

   X2    X4    M1
   ─┤├──┬─┤├──┤/├────────────────────────────────────(Y0      )─  State
   Y0   │                                                         S1
   ─┤├──┘

   X3    X4
   ─┤├──┬─┤├─────────────────────────────────────────(Y1      )─  State
   Y1   │                                                         S2
   ─┤├──┘

   X3    X5
   ─┤├──┬─┤├─────────────────────────────────────────(M1      )─  State
   M1   │                                                         S3
   ─┤├──┘

   M1    X2    X5                                      K50
   ─┤├───┤├───┤├─────────────────────────────────────(T0      )─  State
                                                                  S4

   ─────────────────────────────────────────────────[MCR   N0 ]─

   ─────────────────────────────────────────────────[END      ]─
```

Note

The master contacts are entered as follows:

-[MC N0 M0]-   **-[MCR N0]-**

1. mc   \<space>.   1. mcr   \<space>.
2. n0   \<space>.   2. n0   \<ent>.
3. m0   \<ent>.

Converted ladder diagram – PNEU1

```
       X0     X1     T0   (State S5)
 0    ─┤├────┤├────┤/├──────────────────────────[MC   N0    M0 ]─ Start
       START  STOP                                                Up
       M0
      ─┤├─

 N0    M0

       X2     X4     M1
 7    ─┤├────┤├────┤/├──────────────────────────────────(Y0   )─ State
       A-     B-                                        SOL A    S1
       Y0
      ─┤├─
      SOL A

       X3     X4
12    ─┤├────┤├─────────────────────────────────────────(Y1   )─ State
       A+     B-                                        SOL B    S2
       Y1
      ─┤├─
      SOL B

       X3     X5
16    ─┤├────┤├─────────────────────────────────────────(M1   )─ State
       A+     B+                                                 S3
       M1
      ─┤├─

       M1     X2     X5                                  K50
20    ─┤├────┤├────┤├──────────────────────────────────(T0    )─ State
              A-     B+                                          S4

26    ─────────────────────────────────────────[MCR   N0    ]─

28    ─────────────────────────────────────────────────[END  ]─
```

Note

The serial master contacts N0 ⊣⊢ M0 cannot be programmed in directly. However, they will appear automatically, when the ladder diagram is monitored for the first time.

Principle of operation – single cycle

The following describes the single-cycle operation of PNEU1.

The term 'Piston A operates' or 'Piston B operates' refers to the pistons moving from their back positions, i.e. A– or B–, to their forward positions, A+ or B+.

1. Line 0 (Startup)
   When Input X0 is operated, this will cause the Master Control Instruction -[MC N0 M0]- to be executed. Basically, the master control instruction enables a particular part of the ladder diagram to become operative. This is shown on the ladder diagram as a pair of normally open horizontal contacts, which enable the operation of the instructions from Line 7 to Line 26. If the Instruction -[MC N0 M0]- is not executed, then the instructions from Line 7 to Line 26 will be ignored.
2. Line 7 (State S1)
   With Pistons A and B in the back position, i.e. A– and B–, inputs X2 and X4 will operate, causing Output Y0 to energise and latch over its own normally open contact. This will cause Piston A to move to the A+ position. The Memory Coil M1 will operate later in the cycle, when it is necessary for Piston A to become de-energised. Hence, at this point in the cycle, the normally closed contacts of M1 will remain closed. Even though Input X2 now opens, when Piston A moves forward, the latch circuit ensures that Output Y0 will not become de-energised.
3. Line 12 (State S2)
   With Piston A fully forward, then Input X3 (A+) will operate. This plus Input X4 (B–) will cause Output Y1 to be energised and latch over its own contact. Piston B will now move to the B+ position.
4. Line 16 (State S3)
   At this moment in time, both Pistons will be fully forward and hence Inputs X3 and X5 will be operated. This will cause the Memory Coil M1 to operate and latch over its own contact. The normally closed contact of M1 at Line 7 will now open and break the Y0 latch circuit. This will de-energise Y0 and hence cause Piston A to retract to the A– position.
5. Line 20 (State S4)
   When Piston A returns to the A position, Input X2 will re-make, and this plus M1 and X5 (B+) will energise the Timer Coil T0.
6. Line 0
   After 5 sec, Timer T0 will time out (State S5). The normally closed timer contacts of T0 will open, breaking the start latch circuit. This will cause the Master Contact M0 to open, causing all of the energised outputs from Lines 9–20 to become de-energised. Hence, M1 will become de-energised, as will Output Y1. The de-energising of Output Y1 will cause Piston B to retract to the B– position.

PLC sequence controller 97

7. Line 26

The Instruction [MCR N0] is used to terminate the master control section, and hence, any instructions which follow this instruction will not be affected if the master control is OFF.

8.3 Simulation – PNEU1

Using the simulation unit shown in Figure 8.3, test and monitor the operation of PNEU1.

Simulation unit

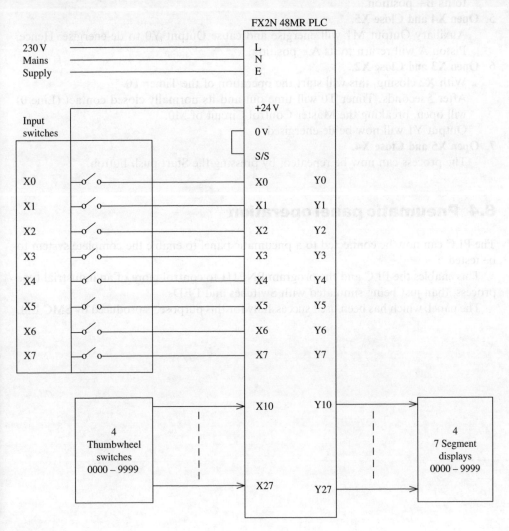

Figure 8.3

Simulation and monitoring procedure

1. **Display and monitor the Ladder Diagram PNEU1.**
2. **Operate the Input Switches X1, X2 and X4.**
   This will simulate the operation of the stop push button and the A− and B−limit switches.
3. **Momentarily operate the Start Switch X0.**
   Output Y0 will now be energised and this will cause Piston A to operate to the A+ position.
   In a real situation with Piston A moving forward, its limit switches X2 would open and X3 would close.
4. **Open X2 and Close X3.**
   This will cause Output Y1 to energise and hence enable Piston B to move forward to its B+ position.
5. **Open X4 and Close X5.**
   Auxiliary Output M1 will energise and cause Output Y0 to de-energise. Hence, Piston A will return to its A− position.
6. **Open X3 and Close X2.**
   With X2 closing, this will start the operation of the Timer T0.
   After 5 seconds, Timer T0 will time out and its normally closed contact (Line 0) will open, breaking the Master Control Circuit of M0.
   Output Y1 will now be de-energised.
7. **Open X5 and Close X4.**
   The process can now be repeated, by pressing the Start push button.

8.4 Pneumatic panel operation

The PLC can now be connected to a pneumatic panel to enable the complete system to be tested.

This enables the PLC and the program PNEU1 to control more of an industrial type process, than just being simulated with Switches and LEDs.

The panel, which has been used successfully for this purpose, is produced by SMC Ltd.

Pneumatic drawing – PNEU1

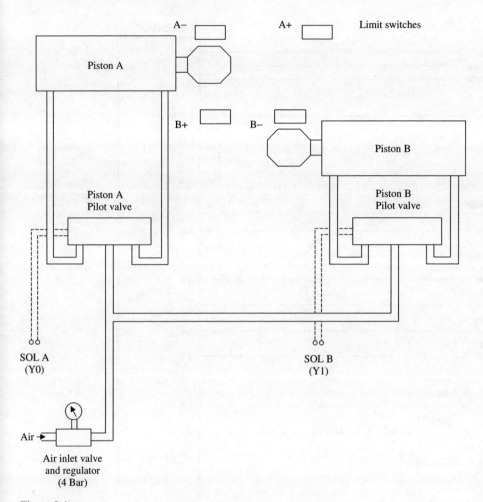

Figure 8.4

PNEU1 wiring diagram (relay output)

The electrical diagram in Figure 8.5, shows how an FX2N PLC system is wired to the SMC pneumatic panel.

This enables the PLC and the program PNEU1 to control an industrial type application, instead of being simulated with input switches and output LEDs.

PNEU1 wiring diagram

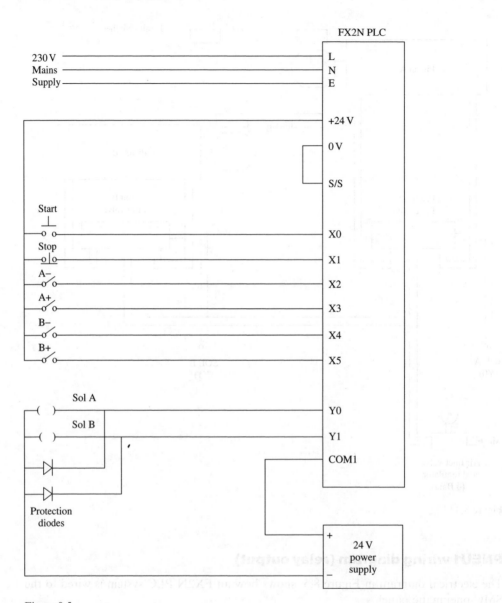

Figure 8.5

8.5 Forced input/output

The Mitsubishi FX2N PLC has the facility to enable inputs and outputs to be turned ON and OFF directly from the computer, without the PLC program running.
This is extremely useful when commissioning a system or when fault finding.
It enables a check to be carried out on the wiring from the inputs and the outputs to the PLC and also, whether or not the input and output devices are operating correctly.

Note

The forcing of inputs when the PLC is in RUN can cause the PLC to automatically start operating. This can cause problems involving safety and should only be done with great care.
The following describes how forcing can be used with the project PNEU1.

Forcing the Start Input X0

1. Ensure that the Ladder Diagram PNEU1 is being displayed and that it has been downloaded to the PLC.
2. Operate the following input switches to simulate the start conditions for PNEU1.
   (a) X1   Stop.
   (b) X2   A−.
   (c) X4   B−.
3. Ensure the PLC is switched to RUN.
4. From the menus select:
   (a) Online.
   (b) Debug.
   (c) Device test.
5. Enter X0, into the Device test window, as shown in Figure 8.6.
6. On selecting FORCE ON, the following will occur:
   (a) The Input X0 will turn ON.
   (b) The Output Y0 will turn ON.
7. The Output Y0 is turned ON due to the Input X0 being forced ON and starting the operation of PNEU1 (Figure 8.7).
8. By following the simulation and monitoring procedure, on page 98, the operation of PNEU1 can be continued.
9. Ensure all forced inputs are now Forced OFF.

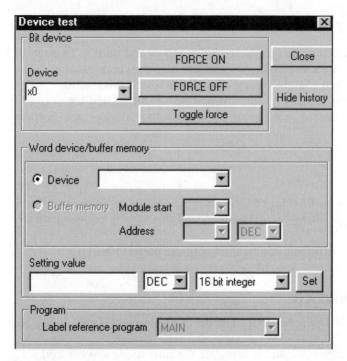

Figure 8.6

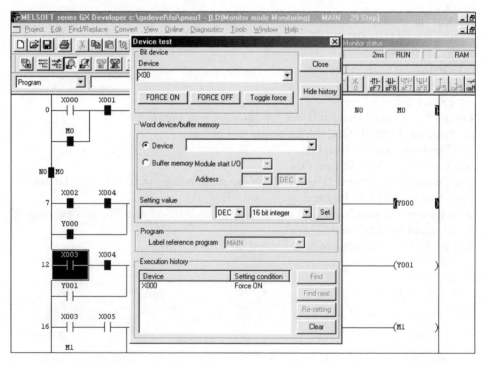

Figure 8.7

Forcing the Y Outputs

The Y Outputs, i.e. Y0 and Y1 can also be forced ON/OFF. However, as the PLC program will override any forced output, it is necessary before forcing outputs to turn the PLC switch from RUN to STOP.

Forcing the Output Y0

1. Ensure that the Ladder Diagram PNEU1 is being displayed and that it has been downloaded to the PLC.
2. Turn the PLC switch from RUN to STOP.
3. Similarly, as for FORCING Inputs, select:
   (a) Online.
   (b) Debug.
4. Select Device test once more.
5. Enter Y0 into the Device window.
6. The Output Y0 can now be either Forced ON and OFF or Toggled ON and OFF.
7. After Forcing Y0 ON, force some of the other outputs ON.
8. Ensure all forced outputs are now forced OFF.

Execution history

1. Check the Execution history display to confirm that all of the outputs, which were forced ON, have now been forced OFF (Figure 8.8).
2. Select Close to return to the ladder diagram.

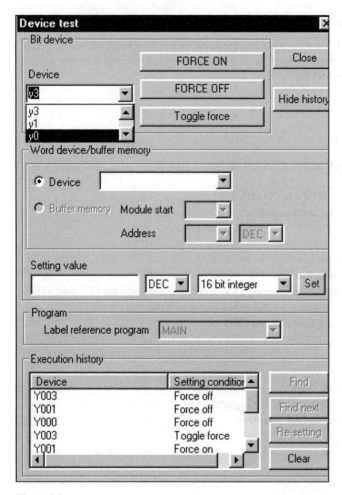

Figure 8.8

8.6 Assignment – PNEU2

Automate PNEU1 so that it will:

1. On operating the start button, carry out three complete cycles before stopping.
2. Repeat the automatic cycle each time the start button is operated.

Note

It is obvious that a counter will be required, hence the following must be considered:

1. What input will be used to enable the counter to count up?
2. What will happen when the counter reaches a count of 3?
3. What input will be used to reset the counter?

9

Free line drawing

The Gx-Developer software has the facility which enables the mouse to draw the connecting lines between the devices in a ladder diagram.

This has many advantages when producing a ladder diagram, which contains parallel connected instructions.

Consider the ladder diagram PNEU1 shown on Page 95.

The procedure of connecting in parallel, each of the normally open contacts Y0, Y1 and M1 with their respective devices is not that easy. However, the use of the free line drawing facility will now be used, to simplify and speed up this task.

1. Create a new project PNEU1A.
2. Select the free-drawn line icon or press F10 (Figure 9.1).

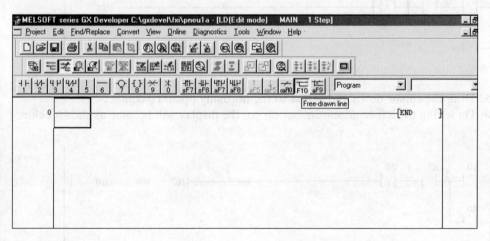

Figure 9.1

3. Enter the ladder diagram as shown below:

```
    X0      X1      T0
 ─┤ ├───┬─┤ ├────┤/├─────────────────────────[MC    N0     M0  ]─
    M0  │
 ─┤ ├───┘

    X2      X4      M1
 ─┤ ├────┤ ├────┤/├──────────────────────────────────────(Y0    )─

    Y0
 ─┤ ├─
```

4. Move the cursor over the normally closed contact M1 and press the left-hand mouse button.

```
    X0      X1      T0
 ─┤ ├───┬─┤ ├────┤/├─────────────────────────[MC    N0     M0  ]─
    M0  │
 ─┤ ├───┘

    X2      X4    ┌─M1─┐
 ─┤ ├────┤ ├─────┤ /├─────────────────────────────────────(Y0    )─
                 └────┘
    Y0
 ─┤ ├─
```

5. Drag the cursor down and across to the normally open Y0 contacts.
6. On letting the left-hand mouse button go, the display will become as shown below:

```
    X0      X1      T0
 ─┤ ├───┬─┤ ├────┤/├─────────────────────────[MC    N0     M0  ]─
    M0  │
 ─┤ ├───┘

    X2      X4      M1
 ─┤ ├────┤ ├────┤/├──────────────────────────────────────(Y0    )─

    Y0    ┌────┐
 ─┤ ├─    │    │
          └────┘
```

7. Complete the remainder of the ladder diagram, as shown below using the free line draw facility.
8. Convert and save PNEU1A.

Ladder diagram – PNEU1A

```
          X0      X1     T0    (State S5)
     0    ┤├──────┤├─────┤/├──────────────────────────[MC   N0    M0  ]
          START   STOP
          M0
          ┤├

   N0  │ M0

          X2      X4     M1
     7    ┤├──────┤├─────┤/├──────────────────────────────────(Y0    )
          A-      B-                                           SOL A
          Y0
          ┤├
          SOL A

          X3      X4
    12    ┤├──────┤├─────────────────────────────────────────(Y1    )
          A+      B-                                           SOL B
          Y1
          ┤├
          SOL B

          X3      X5
    16    ┤├──────┤├──────────────────────────────────────────(M1   )
          A+      B+
          M1
          ┤├

          M1      X2     X5                                    K50
    20    ┤├──────┤├─────┤├──────────────────────────────────(T0    )
                  A-     B+

    26   ──────────────────────────────────────────────────[MCR  N0  ]

    28   ──────────────────────────────────────────────────[END ]
```

9.1 Inserting an output in parallel with an existing output

The free line option can also be used for the paralleling of outputs.

At Line 26 of PNEU1A insert a new line above the Instruction -[MCR N0]-, and using the free line option connect in parallel, the Output Y2 to the existing Timer -(T0 ^{K50})-.

Modified ladder diagram – PNEU1A

Press F4 to convert the modified Ladder Diagram PNEU1A, so that it appears as shown below:

```
        X0   X1   T0
  0    ─┤├──┤├──┤/├──────────────────────────[MC   N0   M0  ]─
        M0
       ─┤├─

        X2   X4   M1
  7    ─┤├──┤├──┤/├──────────────────────────────────(Y0    )─
        Y0
       ─┤├─────────┤

        X3   X4
  12   ─┤├──┤├───────────────────────────────────────(Y1    )─
        Y1
       ─┤├─

        X3   X5
  16   ─┤├──┤├───────────────────────────────────────(M1    )─
        M1
       ─┤├─

        M1   X2   X5                                    K50
  20   ─┤├──┤├──┤├───────────────────────────────────(T0    )─
                                   
                                 ────────────────────(Y2    )─

  27   ─────────────────────────────────────────────[MCR  N0 ]─

  29   ─────────────────────────────────────────────[END    ]─
```

9.2 Delete free line drawing

Using the Delete free-drawn line facility, it is possible to quickly delete any ladder diagram line.

The ladder diagram will now be modified so that at Line 0, the parallel contact M0 will be in parallel with Inputs X0 and X1 (Figure 9.2).

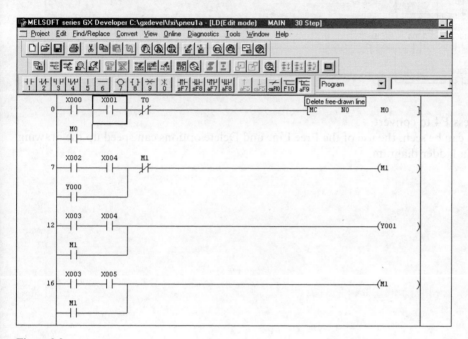

Figure 9.2

1. Select the Delete free-drawn line icon or press the ALT F9 keys.
2. Move the cursor to the X1 contact and then move the cursor down one position to delete the vertical line.

3. Move the cursor now back over the normally closed T0 contact.
4. Select the Free line draw icon or press the F10 key and then holding down the left-hand mouse button move the cursor down one position and then left, to join up with the normally open contact M0.

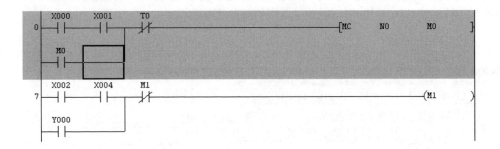

5. Press F4 to convert.
6. As can be seen, the use of the Free Line and Delete options can speed up the drawing of a ladder diagram.

10
Safety

In any industrial design, safety must be a first priority in that the designer(s) must take *all* reasonable steps to ensure a person or persons cannot be harmed in any way whilst the system is being operated.

There are now European Machinery Safety Standards, which also include the control systems of the machinery. These can be summarised as follows:

1. All emergency circuits such as emergency stop buttons and safety guard switches must be hardwired and not depend on software, i.e. PLCs or electronic logic gates.
2. A designer(s) must carry out a risk assessment procedure. In addition the procedure must be documented for possible inspection at any time.
3. Implement any changes to the design to limit any risk, as far as possible.
4. Ensure that where there are remaining risks, the use of safeguards must be implemented into the design.

A company, which specialises in all areas of safety technology including safety relays, is PILZ International.

For further details visit the PILZ website, www.pilz.com.

10.1 Emergency stop requirements

Figure 10.1 represents an emergency stop circuit which is used to protect against a Category 2 fault condition.

With a Category 2 fault condition, the emergency stop circuit must meet the following requirements:

1. It is independent of software and digital logic circuits.
2. The emergency stop button should be wired to a safety relay circuit, i.e. one in which the relays use positive guidance contacts.
3. Have in-built redundancy, in that it is has more than one internal safety relay, e.g. R1 and R2.

10.2 Safety relay specification

To meet the requirements of a Category 2 fault situation, the safety relay must have the following specification:

1. A safety check prior to the system being started up will only allow the system to continue starting up, should no fault be found. This is done as follows:
   (a) Ensure the emergency stop circuit is made;
   (b) A cross-check circuit, i.e. R3, which checks whether R1 or R2 is held stuck in the closed position;
   (c) By using positive guidance contacts, check whether any of the output contacts have become welded together. The positive guidance contacts ensure that none of the normally closed contacts can reclose, if any of the normally open contacts have become welded together.
2. A mains failure facility, which ensures that if there is a mains failure, the outputs cannot be automatically re-energised, when the mains supply is restored.

Safety 113

10.3 Emergency stop circuit – PNEU1

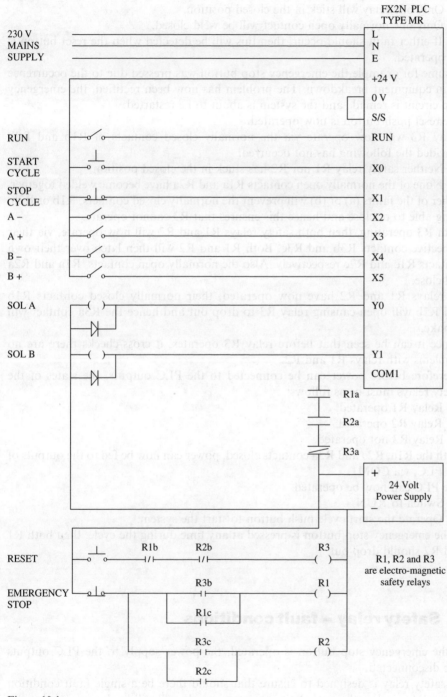

Figure 10.1

Principle of operation

1. This circuit is designed to operate should a single fault occur, i.e.
   (a) Only one relay will stick in the closed position.
   (b) Only one normally open contact will be weld closed.
   (c) If either fault should occur, then this will be detected when the reset button is operated.
2. Assume for example the emergency stop button was pressed due to the occurrence of an equipment breakdown. The problem has now been rectified, the emergency stop circuit is remade and the system is about to be restarted.
3. The reset push button is now operated.
4. Relay R3 will now operate via the normally closed contacts of R1b and R2b provided the following has not occurred:
   (a) Neither safety relay R1 nor R2 has stuck in the closed position.
   (b) None of the normally open contacts R1a and R2a have become welded together. Either of the faults (a) or (b) will prevent the normally closed contacts, R1b or R2b, being able to re-close and hence this ensures that R3 cannot operate.
5. With R3 operating, then both safety relays R1 and R2 will now operate, via their respective contacts R3b and R3c. Both R1 and R2 will then latch over their own contacts R1c and R2c respectively. Also the normally open contacts, R1a and R2a will close.
6. As relays R1 and R2 have now operated, their normally closed contacts R1b and R2b will open causing relay R3 to drop out and hence the R3a contact will remake.
7. Hence it can be seen that before relay R3 operates, it cross-checks there are no problems with relays R1 and R2.
8. Therefore before power can be connected to the PLC outputs, the states of the safety relays must be as follows:
   (a) Relay R1 operated;
   (b) Relay R2 operated;
   (c) Relay R3 not operated.
   With the R1a, R2a and R3a contacts closed, power can now be fed to the outputs of the PLC, via COM1.
9. The PLC can now be operated.
   (a) Switch to RUN;
   (b) Operate the start cycle push button to start the system.
10. If the emergency stop button is pressed at any time during the cycle, then both R1 and R2 should drop out.

10.4 Safety relay – fault conditions

When the emergency stop button is operated, the power supply to the PLC outputs must be disconnected.

The safety relay is designed to ensure that should there be a single fault condition within the safety relay, the output circuit power supply will still be disconnected.

The possible faults within the safety relay are:

1. Sticking relay
2. Welded contacts.

Sticking relay

If one of the relays should stick in, then since this is only a single fault, the other relay will still be able to drop out and hence break the power supply circuit.

Welded contacts

When the emergency stop button is pressed, the circuit is designed to ensure that one relay will remain energised for a few milli-seconds longer than the other.

The effect of this is that if there is a high level of current flowing into an inductive circuit then there will be arcing across the first set of contacts to open, which may cause those contacts to become welded together and for the external current to start flowing again.

When the second set of contacts open a few milli-seconds later, there has been insufficient time for the current to rebuild up to a high enough level in the inductive circuit to cause arcing across these contacts when they open. Therefore, these second set of contacts will open without any problems.

10.5 System start-up check

At the start up of the system after the emergency stop button has been operated or after a mains supply shutdown/failure, the safety relay now carries out an internal check to determine whether or not there is a fault within the safety relay.

1. On operating the reset button, if there has been a fault condition, and one of the relays, either R1 or R2, has stuck in, or one of the R1a or R2a contacts has become welded together, then the respective contact, i.e. R1b or R2b, will not be able to close. Hence when the reset button is pressed, relay R3 will not energise.
2. Therefore, power to the PLC outputs cannot be restored, until the fault condition has been rectified.
3. Should the emergency stop button have been pressed or there has been a mains supply shutdown/failure, then both R1 and R2 will drop out.

When the mains supply is restored, then R1 and R2 will not automatically re-energise, since their latch circuit contacts i.e. R1c and R2c will be open.

Provided there are no fault conditions and the emergency stop circuit is made, then when the Reset button is pressed, relay R3 will re-energise to enable the system to restart.

11

Documentation

The documentation of a ladder diagram is an essential part of the design process. In that without good documentation, those people who are responsible for maintaining any system controlled by a PLC, will have great difficulty in understanding how the system operates.

There are three ways a PLC ladder diagram can be documented, these are:

1. Comments
2. Statements
3. Notes.

1. *Comments*
   A ladder diagram can be commented by associating a meaningful name to a device

i.e.
```
     X0
    ─┤ ├─
    Start
```

The comment Start is associated with the input device X0.

2. *Statements*
   A statement is used:
   (a) At the start of the program to describe its overall purpose.
   (b) At selected points within the program, for example at the start of a section dealing with analogue to digital conversion (ADC).

```
;PNEU1
;Control of a Pneumatic Process

        X0    X1    T0
   0 ───┤├────┤├────┤/├──────────────────────────[MC   N0    M0   ]─
        │                                                          │
        M0                                                         │
        ├┤├─                                                       │
```

3. *Note*
   A note is associated with an output and can be used to describe what the output is being used for in the control process, i.e. Energise Piston A.

```
                                                    *  <Energise Piston A      >
       X2      X4      M1
 7     ┤├──────┤├──────┤/├──────────────────────────────────(Y0    )─
       A-      B-      A&B                                   SOL A
                       FWD

       Y0
       ┤├
```

PNEU1 – comments

The following comments will now be added to the ladder diagram, PNEU1.

Device	Comment
X0	Start
X1	Stop
X2	A–
X3	A+
X4	B–
X5	B+
Y0	Sol A
Y1	Sol B
M0	Start Cycle
M1	A&B FWD
T0	5-sec delay

11.1 Comments

Comments can be displayed on the ladder diagram in one of the following formats:

1. five characters per row up to a maximum of three rows;
2. eight characters per row up to a maximum of two rows;
3. eight characters per row up to a maximum of four rows.

To select the format best suited to a particular application refer to pages 125 and 128.

The format of five characters per row up to a maximum of three rows, will be used throughout this chapter.

For example, the comment for M0 is entered as StartCycle, i.e. without a space between Start and Cycle.

This ensures that when the ladder diagram is displayed with its comments, the comments will be displayed neatly as shown below.

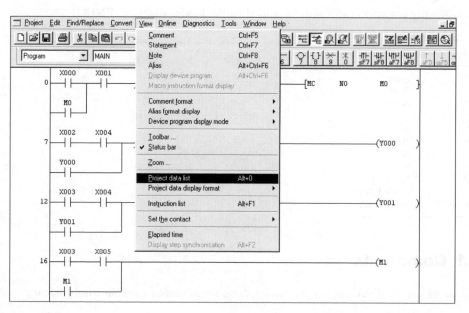

Entering comments

To enter the comments listed on page 117 into the PNEU1 project, the following procedure will be used.

1. Ensure the project data list is displayed on the left of the ladder diagram (see Figure 11.1).

Figure 11.1

2. Alternatively select the project data list icon, shown in Figure 11.2.

Figure 11.2

3. From the project data list, select the following:
   (a) Device comment
   (b) Comment.
4. The display now becomes as shown in Figure 11.3.

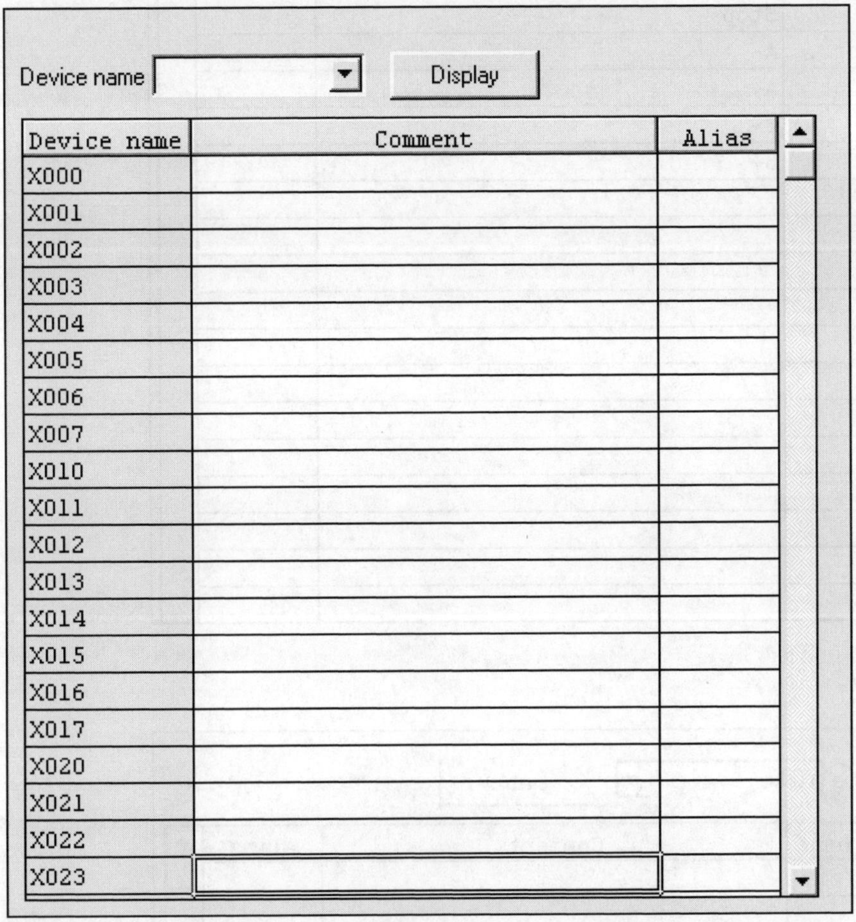
Device name
Display
Device name
Comment
Alias
X000
X001
X002
X003
X004
X005
X006
X007
X010
X011
X012
X013
X014
X015
X016
X017
X020
X021
X022
X023

Figure 11.3

5. (a) Enter X0 into the device window;
   (b) Select display;
   (c) Enter the following comments for X0 to X5 (see Figure 11.4).

Y outputs

To enter the Y output comments, carry out the following.

1. (a) Change the device name to Y0, as shown in Figure 11.5;
   (b) Select display.

Device name X0 Display

Device name	Comment	Alias
X000	Start	
X001	Stop	
X002	A-	
X003	A+	
X004	B-	
X005	B+	
X006		
X007		
X010		
X011		
X012		
X013		
X014		
X015		
X016		
X017		
X020		
X021		

Figure 11.4

Device name Y000 Display

Device name	Comment	Alias
Y000		
Y001		
Y002		
Y003		
Y004		
Y005		
Y006		
Y007		
Y010		
Y011		
Y012		

Figure 11.5

2. Enter the Y0 and Y1 output comments.
   (a) Sol A
   (b) Sol B (see Figure 11.6).

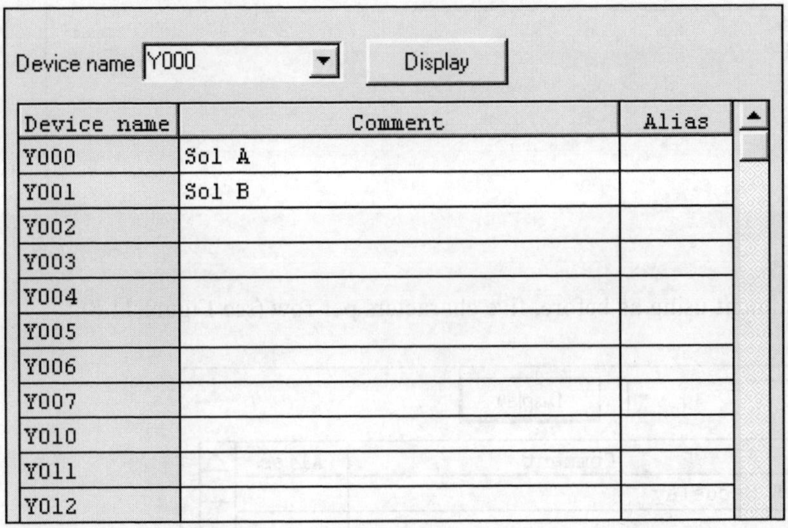

Device name Y000 Display

Device name	Comment	Alias
Y000	Sol A	
Y001	Sol B	
Y002		
Y003		
Y004		
Y005		
Y006		
Y007		
Y010		
Y011		
Y012		

Figure 11.6

M coil comments

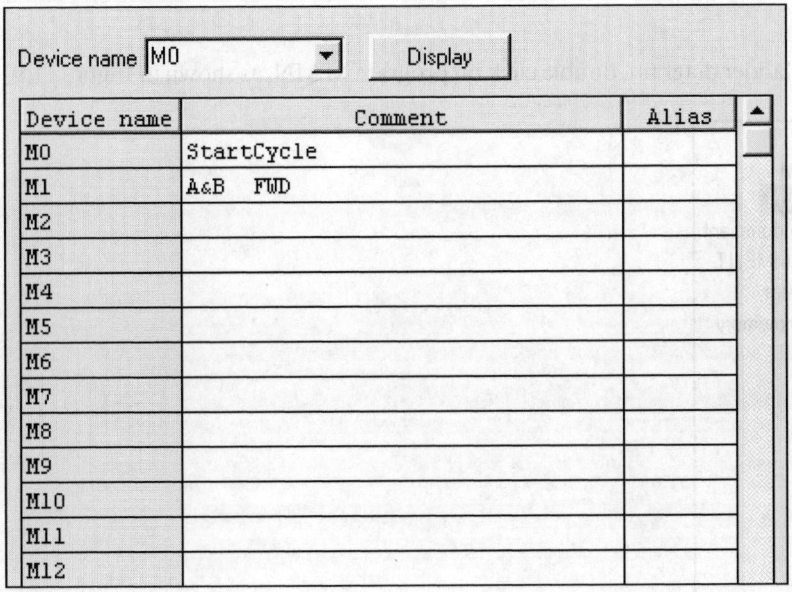

Device name M0 Display

Device name	Comment	Alias
M0	StartCycle	
M1	A&B FWD	
M2		
M3		
M4		
M5		
M6		
M7		
M8		
M9		
M10		
M11		
M12		

Figure 11.7

Note: Enter the comments exactly as shown in Figure 11.7.

This will ensure that the M0 comment for example, will be displayed on the ladder diagram as shown below.

Timer comments

Enter the T0 comment using as before, five characters per row (see Figure 11.8).

Device name	Comment	Alias
T0	5 Secdelay	
T1		
T2		
T3		
T4		

Device name T0   Display

Figure 11.8

To return to the ladder diagram, double click on program MAIN, as shown in Figure 11.9.

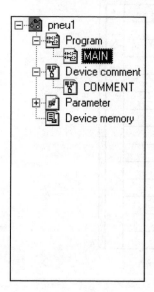

Figure 11.9

11.2 Statements

The following statements will now be entered, using the procedure described on the following pages.

(a) Line 0
   PNEU1
   Control of Pneumatic System
(b) Line 20
   Timer Section

Entering the statements

To enter the statements carry out the following:

1. Press F2 to ensure the ladder diagram is in Write mode.
2. Select Edit.
3. Select Documentation.
4. Select Statement.
5. The display will then return to the ladder diagram.
6. Double click on the input X0, where the line statement is to be positioned.
7. Within the line statement window, enter PNEU1 <ent> (see Figure 11.10).

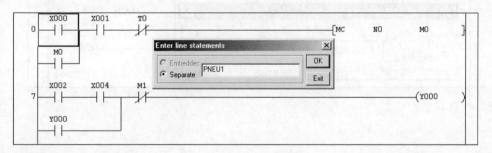

Figure 11.10

8. *Note*
   (a) The term 'separate' indicates that the statement cannot be downloaded and saved in the PLC.
   (b) The term 'Embedded' indicates that if this option could be selected, then statements could be downloaded and saved within the PLC. However this function is only available with the Q Series range of PLCs.

124 *Mitsubishi FX Programmable Logic Controllers*

9. Press F4 to convert the statement so it becomes integrated with the ladder diagram.
10. Double click on X0 again and enter the following statement:
    Control of Pneumatic System <ent> (see Figure 11.11).

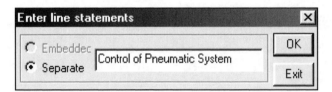

Figure 11.11

11. Press F4.
12. At Line 20 double click on M1 and enter the statement Timer Section <ent> (see Figure 11.12).

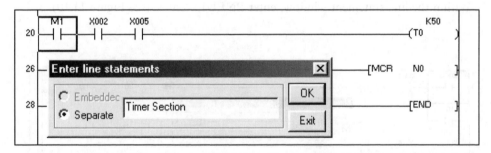

Figure 11.12

13. Press F4.

11.3 Display of comments and statements

The following describes how the comments and statements can now be displayed.

1. Ensure the ladder diagram PNEU1 is being displayed.
2. From the main menu, select View.
3. Select Comment (see Figure 11.13).

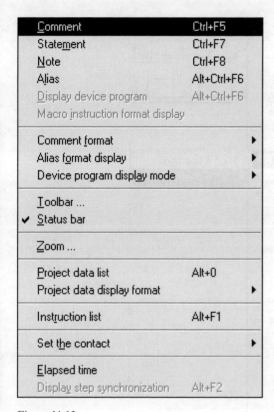

Figure 11.13

4. Select the main menu once more but this time, select Statement.
5. Selecting comment and statement as described above, will now enable the comments and statements to be displayed on the ladder diagram.

11.4 Comment display – 15/16 character format

To ensure the comments are displayed in rows of five characters per row, carry out the following.

1. Select View.
2. Select Comment format.
3. Select 3 * 5 characters.
4. This indicates that the comments can be displayed in rows of five characters with a total of three rows, i.e. a maximum total of 15 characters per comment (see Figure 11.14).

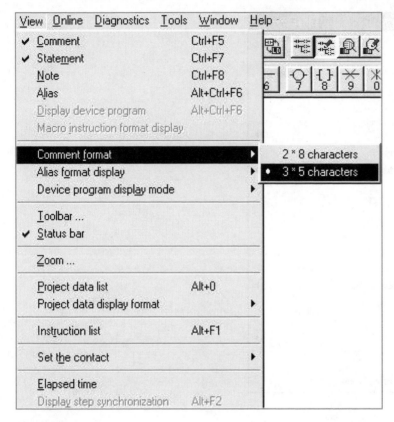

Figure 11.14

5. Alternatively, the comments can be displayed in either two rows of eight characters per row, i.e. a total of 16 characters, or four rows of eight characters per row i.e. a total of 32 characters.
6. The procedure for obtaining either a total of 16 characters or 32 characters is described on page 128.
7. The display will now return to the ladder diagram with the comments and statements now displayed.

PNEU1 – comments and statements

- PNEU1
- Control of Pneumatic System

```
          X0     X1     T0
     0    ─┤├────┤├────┤/├──────────────────────────────[MC   N0    M0   ]─
          Start  Stop  5 Sec                                        Start
                       Delay                                        Cycle

          M0
          ─┤├─
          Start
          Cycle


    N0    M0
    ───────┬─────
           │
           │
           │
          X2     X4     M1
     7    ─┤├────┤├────┤/├──────────────────────────────────────(Y0   )─
          A-     B-     A&B                                       Sol A
                        FWD

          Y0
          ─┤├─
          Sol A


          X3     X4
    12    ─┤├────┤├──────────────────────────────────────────────(Y1   )─
          A+     B-                                                Sol B

          Y1
          ─┤├─
          Sol B
```

(Continued)

```
       X3     X5
    ┌──┤ ├───┤ ├──────────────────────────────────────( M1  )─
 16 │   A+    B+                                        A&B
    │                                                   FWD
    │
    │   M1
    └──┤ ├──
        A&B
        FWD


 *  Timer Section
        M1     X2     X5                                 K50
 20 ───┤ ├───┤ ├───┤ ├──────────────────────────────( T0    )─
        A&B    A-     B+                              5 Sec
        FWD                                           Delay



 26 ──────────────────────────────────────────────[ MCR  N0  ]─



 28 ─────────────────────────────────────────────────[ END ]─
```

11.5 Comment display – 32 character format

In very large projects, where there are thousands of inputs and outputs, the use of only 15 or 16 characters to describe all of the I/O may be insufficient.

In such circumstances the use of 32 characters for each comment may be necessary. To obtain this facility the following set up has to be carried out.

1. From the Tools menu select Options.
2. The display now appears as shown in Figure 11.15.
3. Select Whole data (see Figure 11.16).
4. Note that the Device comment edit/show no., i.e. the total number of characters allocated to each comment is 16.
5. Change this to 32.
6. Select OK.
7. From the main menu, select View.
8. On selecting the Comment format once more as shown in Figure 11.17, it can be seen that the format can now be either of the following:
   (a) 4 × 8 characters
   (b) 3 × 5 characters.

Documentation 129

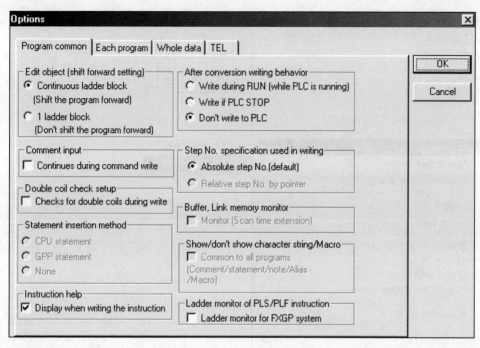

Figure 11.15

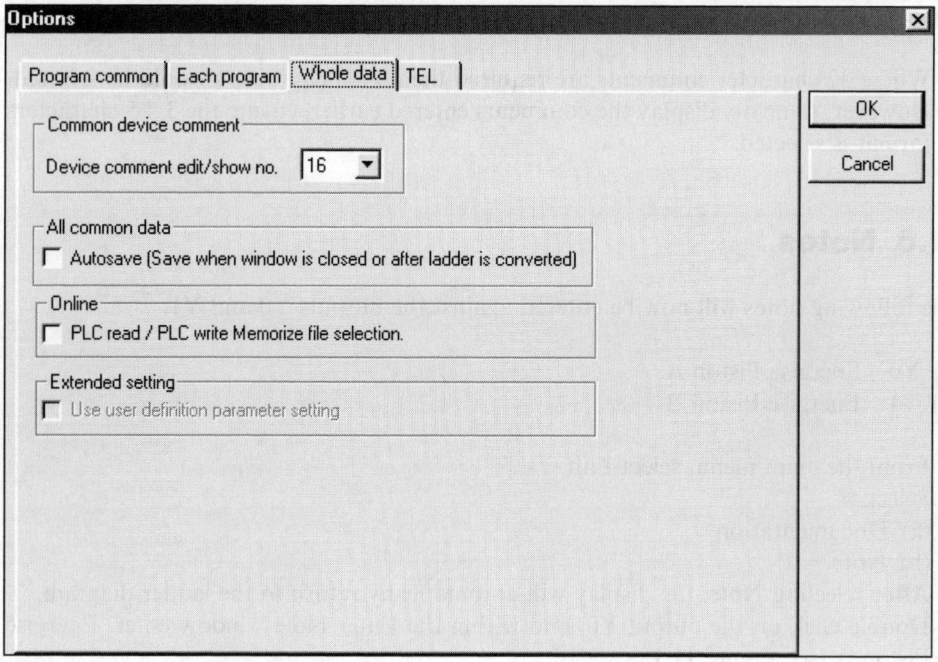

Figure 11.16

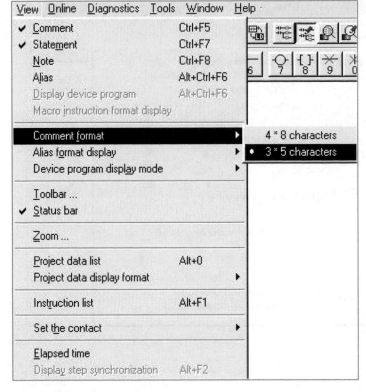

Figure 11.17

9. Where 32 character comments are required then 4 * 8 characters would be selected. However, to neatly display the comments entered earlier, ensure the 3 * 5 characters format is selected.

11.6 Notes

The following notes will now be entered against the outputs Y0 and Y1.

(a) Y0   Energise Piston A
(b) Y1   Energise Piston B.

1. From the main menu, select Edit.
2. Select:
   (a) Documentation
   (b) Note.
3. After selecting Note, the display will automatically return to the ladder diagram.
4. Double click on the output Y0, and within the Enter Note window enter, Energise Piston A (see Figure 11.18).
5. Select OK to return to the Ladder Diagram.

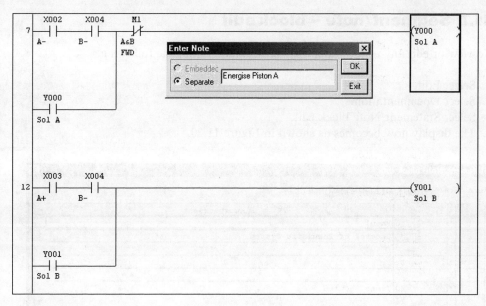

Figure 11.18

6. Press F4.
7. Repeat the procedure to enter the following note with the output Y1.
   (a) Double click on output Y1
   (b) Energise Piston B (see Figure 11.19).
8. Select OK to return to the ladder diagram.
9. Press F4.

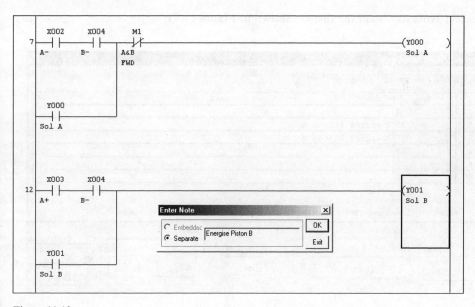

Figure 11.19

11.7 Segment/note – block edit

To view or edit any of the statements or notes, carry out the following.

1. Select Edit.
2. Select Documentation.
3. Select Statement/Note block edit.
4. The display now becomes as shown in Figure 11.20.

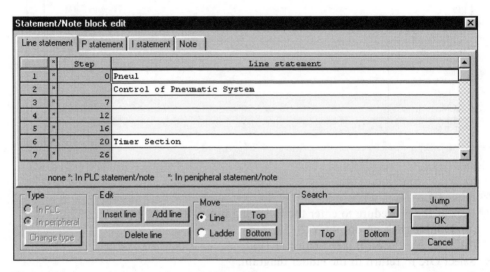

Figure 11.20

5. Select Note to obtain the display shown in Figure 11.21.

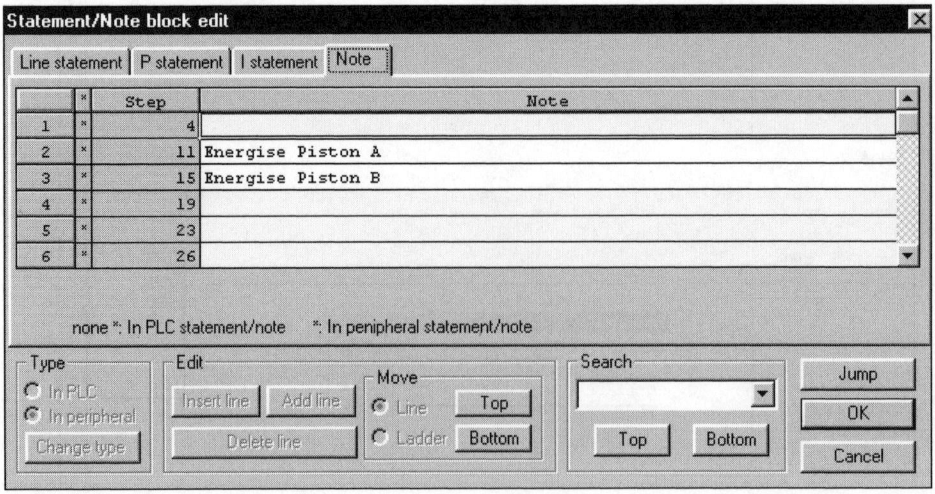

Figure 11.21

11.8 Ladder diagram search using statements

Where a project has been fully documented, i.e. statements have been used at the start of each major section in the ladder diagram, then the statement/note block edit function can be used to jump to the start of the section, where the statement is displayed.

This can be extremely useful for fault finding, in that if problems are occurring at a known section of the ladder diagram then the statement at the start of that section can be selected and by then selecting jump, the ladder diagram starting at that section will be displayed.

1. Ensure that the line statements are being displayed.
2. Select the line statement number, i.e. no. 6 for the Timer Section.
3. Line 6, step 20 now becomes highlighted, as shown in Figure 11.22.

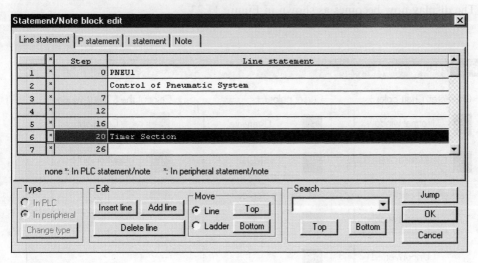

Figure 11.22

4. Select Jump.
5. Select OK.
6. The ladder diagram now appears as shown below, starting at line 20.

```
Timer Section
         M1    X2    X5                                                      K50
   20   ─┤ ├──┤ ├──┤ ├─────────────────────────────────────────────────────(T0     )─
         A&B   A-    B+                                                      5 SEC
         FWD                                                                 DELAY


   26   ──────────────────────────────────────────────────────────────────[MCR  N0 ]─



   28   ──────────────────────────────────────────────────────────────────────[END ]─
```

134   *Mitsubishi FX Programmable Logic Controllers*

7. Repeat the process to jump to the start of the ladder diagram by selecting the statement PNEU1.

11.9 Change of colour display

When displaying comments, statements and notes on the ladder diagram, the default colour is green. To change them all to a different colour i.e. black, carry out the following:

1. Ensure the ladder diagram PNEU1 is being displayed.
2. From the main menu, select Tools.
3. Select Change display colour.
4. The display now becomes as shown in Figure 11.23.

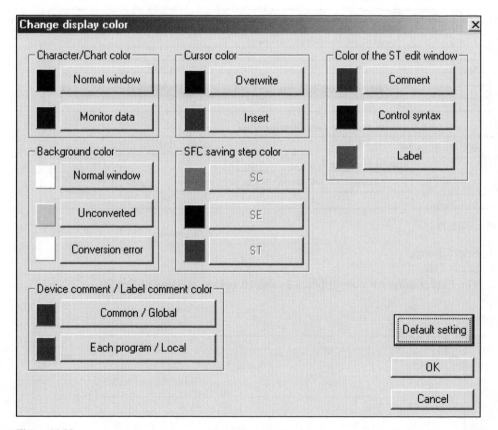

Figure 11.23

5. Select Common/Global and from the colour palette, which is now displayed, select black.

6. Select OK to return to the above display.
7. Select OK once more, to remove the Change display colour menu.

11.10 Display of comments, statements and notes

The following describes how comments, statements and notes can be displayed (see Figure 11.24).

1. Ensure the ladder diagram PNEU1 is being displayed.
2. From the main menu, select View.
3. Select Comment.
4. Repeat the procedure for:
    (a) Statement
    (b) Note.

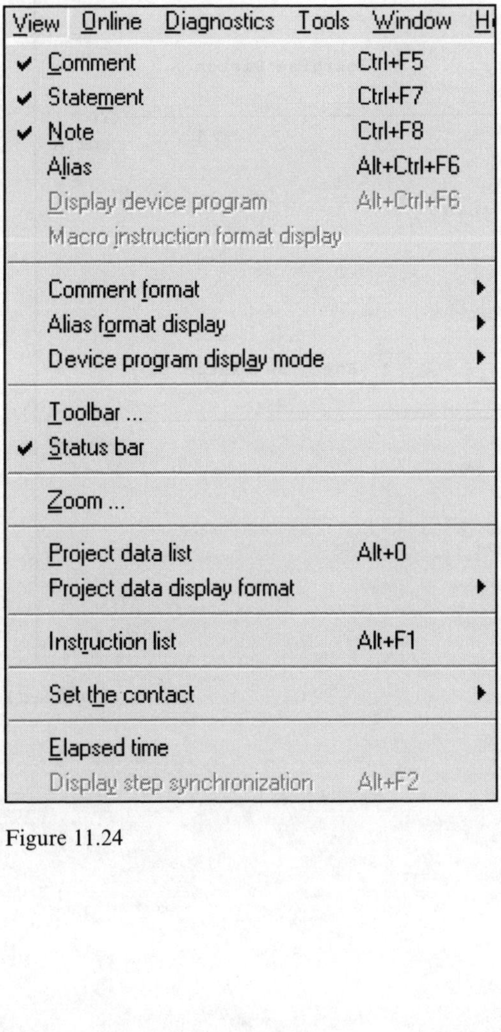

View	Online	Diagnostics	Tools	Window	H
✔ Comment				Ctrl+F5	
✔ Statement				Ctrl+F7	
✔ Note				Ctrl+F8	
Alias				Alt+Ctrl+F6	
Display device program				Alt+Ctrl+F6	
Macro instruction format display					
Comment format				▶	
Alias format display				▶	
Device program display mode				▶	
Toolbar ...					
✔ Status bar					
Zoom ...					
Project data list				Alt+0	
Project data display format				▶	
Instruction list				Alt+F1	
Set the contact				▶	
Elapsed time					
Display step synchronization				Alt+F2	

Figure 11.24

PNEU1 – comments, statements and notes

- PNEU1
- Control of Pneumatic System

```
       X0      X1      T0
   0 ──┤ ├────┤ ├────┤/├──────────────────────────────[MC    N0    M0 ]
      Start  Stop   5 Sec                                         Start
                    Delay                                         Cycle

       M0
      ─┤ ├─
      Start
      Cycle


N0 ──┤M0├──

                                                * <Energise Piston A      >
       X2      X4      M1
   7 ──┤ ├────┤ ├────┤/├──────────────────────────────────────(Y0    )
       A-      B-     A&B                                        Sol A
                      FWD

       Y0
      ─┤ ├─
      Sol A


                                                * <Energise Piston B      >
       X3      X4
  12 ──┤ ├────┤ ├─────────────────────────────────────────────(Y1    )
       A+      B-                                                Sol B

       Y1
      ─┤ ├─
      Sol B
```

(Continued)

```
          X3    X5                                    (M1  )
    16    ─┤├───┤├────────────────────────────────────
          A+    B+                                    A&B
                                                      FWD


          M1
         ─┤├──┘
          A&B
          FWD

  * Timer Section
          M1    X2    X5                              K50
    20   ─┤├───┤├────┤├──────────────────────────────(T0  )
          A&B   A-    B+                              5 Sec
          FWD                                         Delay


    26   ──────────────────────────────────────[MCR  N0  ]


    28   ─────────────────────────────────────────────[END ]
```

11.11 Printouts

This section describes how to obtain a printout of the following:

- Title page
- Header
- Ladder diagram
- Parameters
- Page numbers.

1. Select the following:
   (a) Project
   (b) Print.
2. If necessary modify the display so that it appears as shown in Figure 11.25.

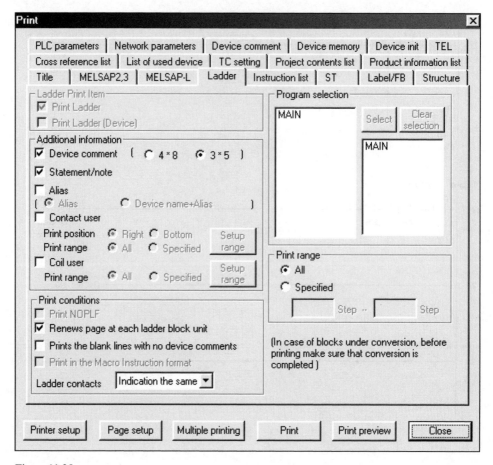

Figure 11.25

Title page

1. To obtain a title page, select Title from the upper print menu.
2. Enter some meaningful text, as shown in Figure 11.26.
3. Select Ladder from the upper print menu, to return to the main print menu.

Header

The header is text that can be printed at the top of every page, which is printed out.

1. From the lower print menu, select Page setup.
2. The display now becomes as shown in Figure 11.27.

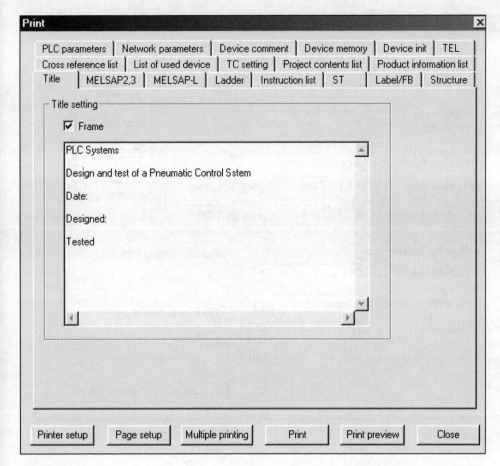

Figure 11.26

3. Select Edit header and enter a one-line description of the project i.e. Pneumatic Controller-PNEU1 (see Figure 11.28).
4. Select OK to return to the page setup menu.

Page numbers

1. To obtain page numbers select Page setup, as shown on page 138.
2. Ensure the Page No. setting is as shown in Figure 11.29.
3. Select OK to return to the main print menu.

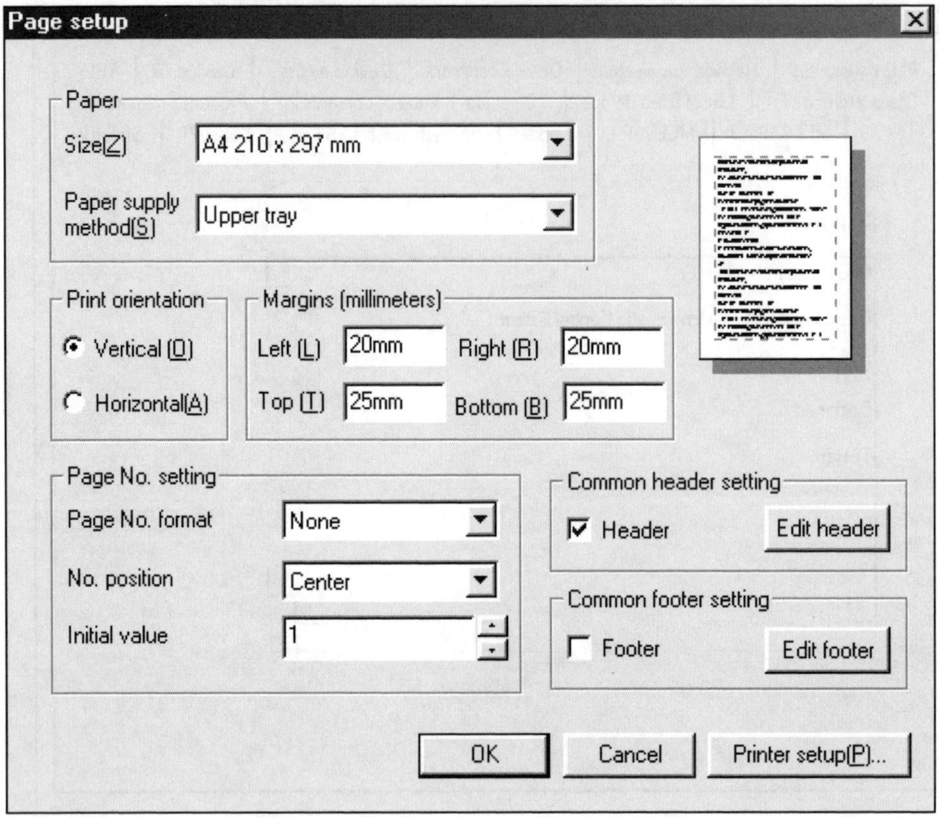

Figure 11.27

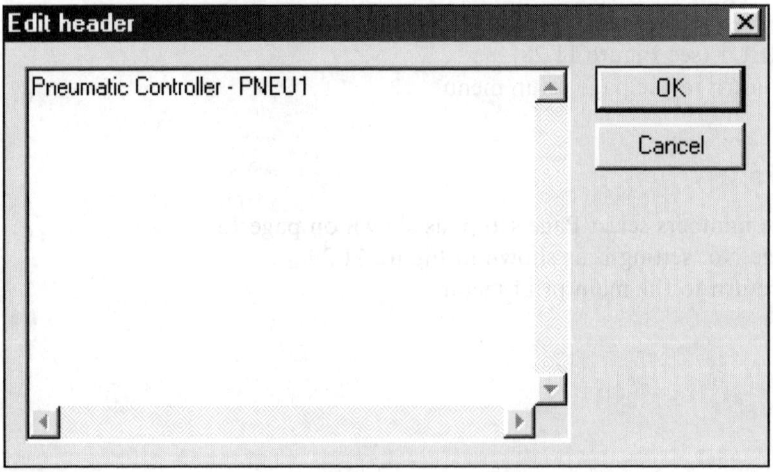

Figure 11.28

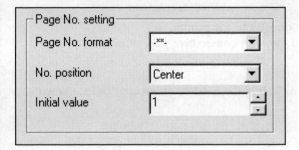

Figure 11.29

11.12 Multiple printing

1. To obtain the printouts it is now necessary to select Multiple print and tick the following as shown in Figure 11.30.
   (a) Title
   (b) PLC parameter
   (c) Ladder.

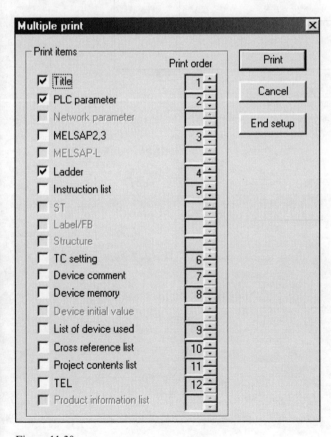

Figure 11.30

2. Select Print.
3. Select Yes, to send the print data directly to the printer.
4. The following will now be printed out with consecutive page numbers.
    (a) The title page
    (b) The parameters
    (c) The ladder diagram PNEU1 including the header, comments, and statements.

Title page

```
Pneumatic Controller – PNEU1

    PLC Systems

    Design and test of a Pneumatic Control System

    Date

    Designed

    Tested
```

PLC parameters list – 1

Pneumatic Controller – PNEU1

PLC name set		
Title	[	]

Mem capcty set

1	Memory size	[8000]
2	Program size	[7000] step
3	Comment size	[2] block [1000] pt
4	File register	[0] block [0] pt

PLC system set

1	Batteryless mode	<OFF>
2	Modem Initialization	<None>
3	Run Contact	× [000]
4	Communication setting	
	Protocol	
	Data length	
	Parity	
	Stop bit	
	Baud rate	
	Header	
	Terminator	
	Control line	
	H/W type	
	Control mode	
	Sum check	
	Trans control Proc	
	StationNo. Setting	[ ] H
	Time out	[ ] × 10ms

Note: The Program size has been reduced from 8000 steps to 7000 steps, for as will be shown later 1000 points, i.e. the equivalent of 1000 steps will be allocated for saving Comments within the PLC. Refer page 146.

PLC parameters list – 2

Pneumatic Controller – PNEU1

Device set

Device name	Sym	Device pt	Device range	Latch range
Internal relay	M	[3072]pt	[0]--[3071]	[500]--[1023]
State	S	[1000]pt	[0]--[999]	[500]--[999]
Timer	T	[256]pt	[0]--[255]	
Counter(16bit)	C	[200]pt	[0]--[199]	[100]--[199]
Counter(32bit)	C	[56]pt	[200]--[255]	[220]--[255]
Data register	D	[8000]pt	[0]--[7999]	[200]--[511]
RAM File register	D			

I/O set

Device name	Sym	Device pt	Device range
Input relay	X	[256]pt	[000]--[377]
Output relay	Y	[256]pt	[000]--[377]

Pneumatic Controller – PNEU1

```
:PNEU1
:Control of Pneumatic System

          X0      X1      T0
  0       ┤├──────┤├──────┤/├────────────────────────────[MC   N0     M0   ]
          Start   Stop    5 Sec                                       Start
                          Delay                                       Cycle
          M0
          ┤├─┐
          Start
          Cycle

   N0   │ M0
        │ ┤├─
          Start
          Cycle

          X2      X4      M1                    * < Energise Piston A       >
  7       ┤├──────┤├──────┤/├──────────────────────────────────────────(Y0  )
          A-      B-      A&B                                          Sol A
                          FWD
          Y0
          ┤├─┐
          Sol A

          X3      X4                            * < Energise Piston B       >
 12       ┤├──────┤├──────────────────────────────────────────────────(Y1   )
          A+      B-                                                   Sol B
          Y1
          ┤├─┐
          Sol B

          X3      X5
 16       ┤├──────┤├──────────────────────────────────────────────────(M1   )
          A+      B+                                                   A&B
                                                                       FWD
          M1
          ┤├─┐
          A&B
          FWD

;Timer Section
          M1      X2      X5                                           K50
 20       ┤├──────┤├──────┤├──────────────────────────────────────────(T0   )
          A&B     A-      B+                                           5 Sec
          FWD                                                          Delay

 26       ├───────────────────────────────────────────────────[MCR  N0   ]─

 29       ├───────────────────────────────────────────────────────[END   ]─
```

146 *Mitsubishi FX Programmable Logic Controllers*

11.13 Saving comments in the PLC

Documentation as mentioned previously is essential for easier fault finding, therefore by saving the comments within the PLC itself, they are immediately available when the program is uploaded.

This is obviously a very useful feature, if for some reason the fully documented master disk cannot be obtained.

The main disadvantage with saving comments within the PLC is that they require being stored in the RAM area of the PLC and hence they reduce the memory available for program use.

For example if 1000 points, i.e. 2000 bytes of memory is reserved for storing comments, then the number of steps available for the program falls to 7000.
Refer Note of page 143 PLC parameters list – 1.
Note: With the FX range of PLCs, statements and notes cannot be stored in the PLC.

Procedure for storing comments

1. Obtain the project data list by selecting:
   (a) View
   (b) Project data list.
2. The project data list now appears on the left hand side of the display as shown in Figure 11.31.

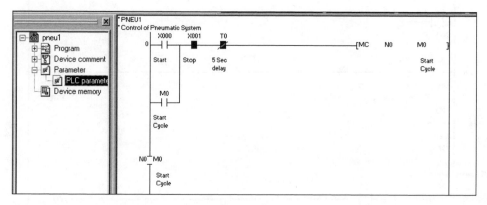

Figure 11.31

3. Double click on PLC parameter to obtain the FX parameter display.
4. Ensure the option Memory capacity has been selected.
5. Reserve two blocks of memory for storing comments (see Figure 11.32).

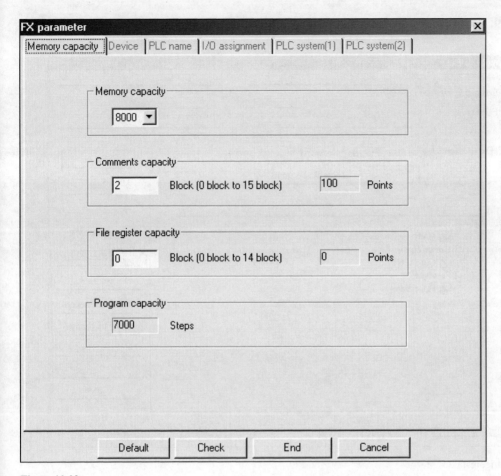

Figure 11.32

6. *Note*
   (a) There is an error in the display in that the allocation of two blocks of comment capacity corresponds to 1000 points and not 100 points. (refer PLC parameters, page 143).
   (b) The program capacity has been reduced from 8000 steps to 7000 steps.
7. Select End to return to the ladder diagram and project data list display.

Downloading the comments to the PLC

To download the ladder diagram and the comments to the PLC, carry out the following:

1. Select Online.
2. Select Write to PLC.
3. Click on Select all, so that COMMENT is now included as part of the download selection.

4. The display now becomes as shown in Figure 11.33.

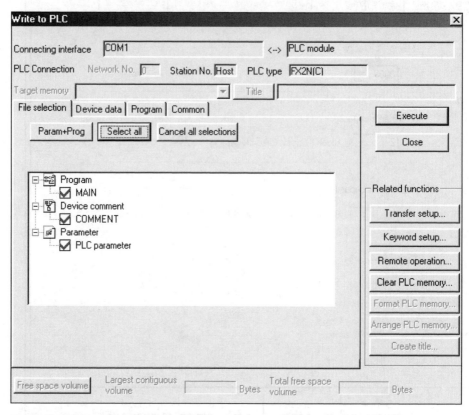

Figure 11.33

5. Select Execute.
6. Select Yes for the program, the comments and the parameters to be downloaded to the PLC.

Uploading the program and comments

It is advisable when uploading a program that it is uploaded into a new project.

This ensures that should there be any differences between the two programs, the program on the master disk does not get overwritten.

References
   Verification   page 62.
   Uploading a project from a PLC   page 63.

1. Open a new project with the name PNEU1B.
2. Select Online.
3. Select Read from PLC.

4. Ensure the following have been selected.
   (a) Main
   (b) Comment
   (c) PLC parameter (see Figure 11.34).

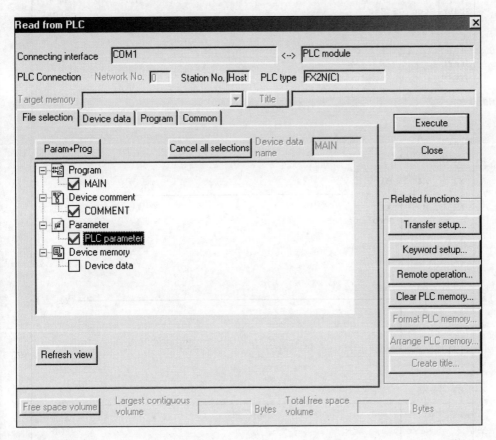

Figure 11.34

5. Select Execute to upload both the ladder diagram and the comments.
6. *Note*: The comments format will be identical to that of PNEU1, i.e. 3×5 characters (refer page 125).

Ladder diagram – PNEU1B

```
     X0    X1    T0
 0   ┤├────┤├────┤/├─────────────────────────────[MC   N0    M0   ]
   Start  Stop  5 Sec                                        Start
              Delay                                          Cycle
     M0
     ┤├
   Start
   Cycle

N0 ── M0
      Start
      Cycle

     X2    X4    M1
 7   ┤├────┤├────┤/├─────────────────────────────────────(Y0    )
     A-    B-    A&B                                     Sol A
                FWD
     Y0
     ┤├
   Sol A

     X3    X4
12   ┤├────┤├─────────────────────────────────────────────(Y1    )
     A+    B-                                              Sol B
     Y1
     ┤├
   Sol B

     X3    X5
16   ┤├────┤├─────────────────────────────────────────────(M1    )
     A+    B+                                              A&B
                                                          FWD
     M1
     ┤├
   A&B
   FWD

     M1    X2    X5                                         K50
20   ┤├────┤├────┤├──────────────────────────────────────(T0    )
     A&B   A-    B+                                        5 Sec
     FWD                                                  Delay

26   ├─────────────────────────────────────────────────[MCR   N0   ]

28   ├──────────────────────────────────────────────────────[END ]
```

150 *Mitsubishi FX Programmable Logic Controllers*

12

Entry ladder monitoring

Entry ladder monitoring is a method, which enables individual rungs in a ladder diagram to be selected and monitored.

When fault finding, entry ladder monitoring is an extremely useful facility if the rungs of the PLC program, which is controlling the operation of the process/machine, are not grouped together in recognisable sections.

The selected rungs, which control the area where the fault occurred, can be displayed and monitored on one screen and therefore the nature of the fault can be more quickly determined and corrected.

To demonstrate the operation of entry ladder monitoring, the ladder diagram PNEU1 will be used.

12.1 Ladder diagram – PNEU1

```
:PNEU1
:Control of Pneumatic System

         X0      X1      T0
    0   ─┤├─────┤├─────┤/├──────────────────────────────[MC   N0    M0    ]─
         Start  Stop   5 Sec                                        Start
                       Delay                                        Cycle
         M0
        ─┤├─
         Start
         Cycle

    N0 │ M0
       ├──┤├
       │  Start
       │  Cycle
       │
       │     X2      X4      M1                   * <  Energise Piston A  >
       │ 7  ─┤├─────┤├─────┤/├──────────────────────────────────────(Y0   )─
       │     A-      B-      A&B                                    Sol A
       │                     FWD
       │     Y0
       │    ─┤├─
       │     Sol A
       │
       │     X3      X4                           * <  Energise Piston B  >
       │ 12 ─┤├─────┤├──────────────────────────────────────────────(Y1   )─
       │     A+      B-                                              Sol B
       │     Y1
       │    ─┤├─
       │     Sol B
       │
       │     X3      X5
       │ 16 ─┤├─────┤├──────────────────────────────────────────────(M1   )─
       │     A+      B+                                              A&B
       │                                                             FWD
       │     M1
       │    ─┤├─
       │     A&B
       │     FWD
       │
       │;Timer Section
       │     M1      X2      X5                                     K50
       │ 20 ─┤├─────┤├─────┤├───────────────────────────────────────(T0   )─
       │     A&B     A-      B+                                     5 Sec
       │     FWD                                                    Delay
       │
    26 ├────────────────────────────────────────────────────────[MCR  N0   ]─


    28 ├─────────────────────────────────────────────────────────────[END ]─
```

12.2 Principle of operation – entry ladder monitoring

To demonstrate the operation of entry ladder monitoring, lines 0 and 20 will be selected.

1. Place the mouse cursor at the start of the statement PNEU1.
2. Hold the left-hand mouse button down and move the cursor slightly downwards.
3. The two statements and all of line 0 will now be highlighted in blue (see Figure 12.1).

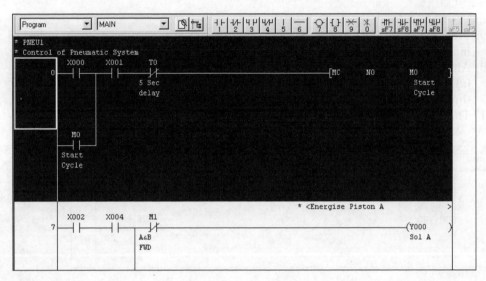

Figure 12.1

4. Press the keys CTRL + C to copy this section of the program to the clipboard.
5. Alternatively select the following:
    (a) Edit
    (b) Copy.
6. Select the following:
    (a) Online
    (b) Monitor
    (c) Entry ladder monitor.
7. Alternatively select the entry ladder monitor icon shown in Figure 12.2.

Figure 12.2

8. The display now becomes as shown in Figure 12.3.

154 *Mitsubishi FX Programmable Logic Controllers*

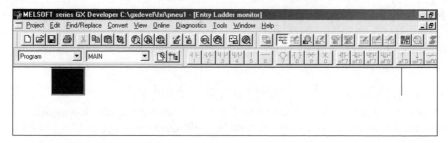

Figure 12.3

9. Press the keys CTRL + V to paste the first section of the ladder diagram at line 0 into the entry ladder monitor.
10. Alternatively select from the top of the display:
    (a) Edit
    (b) Paste.
11. The display now becomes as shown below.

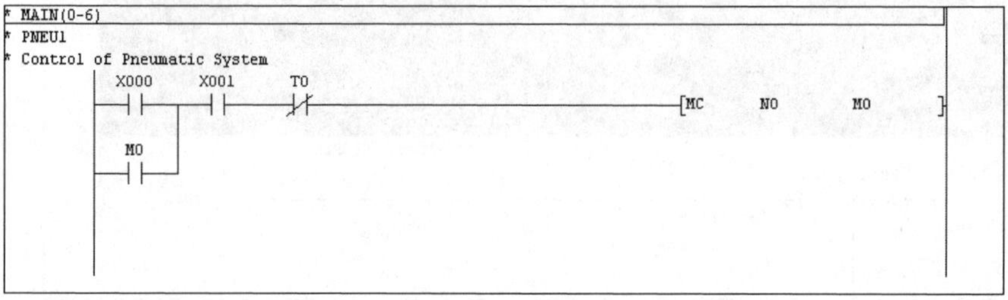

12. The statement Main (0–6) indicates that steps 0–6 have been copied from the main program to the entry ladder monitor display.
13. To return to the main ladder diagram select Main as shown in Figure 12.4.

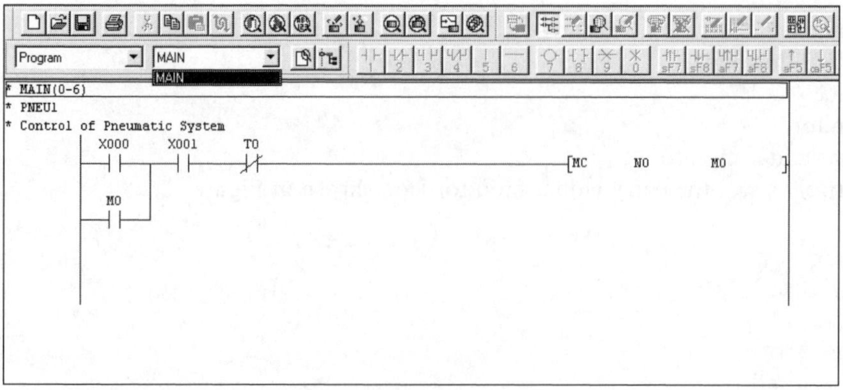

Figure 12.4

14. Repeat the Copy and Paste procedure to copy the timer section at line 20 to the entry ladder monitor diagram (see Figure 12.5).

Figure 12.5

15. The entry ladder monitor display now becomes as shown below.

```
* MAIN(0-6)
* PNEU1
* Control of Pneumatic System
        X000    X001    T0
         ┤├──────┤├─────┤/├────────────────────[MC    N0    M0 ]
        M0
         ┤├

* MAIN(20-25)
* Timer Section
        M1      X002    X005                                K50
         ┤├──────┤├──────┤├─────────────────────────────────(T0   )
```

16. Press F3 to start monitoring.
17. Operate the following input switches to simulate the operation of PNEU1.
    (a) X1   Stop
    (b) X2   A−
    (c) X4   B−.
18. Momentarily operate X0 to start the pneumatic operation by energising Y0.
19. Carry out the remainder of the pneumatic operation until timer T0 is energised.
    (a) Open X2 and close X3 − this will cause Y1 to energise.
    (b) Open X4 and close X5 − this will cause Y0 to de-energise.
    (c) Open X3 and close X2 − this will start the operation of timer T0.

20. While timer T0 is timing out, the display will be as shown below.

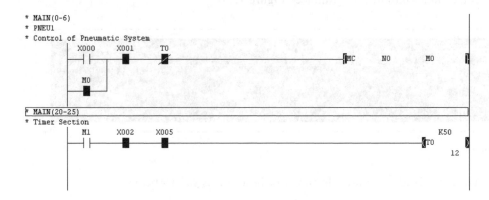

21. After five seconds have elapsed, timer T0 will time out and its normally closed contacts will break the master control circuit.

12.3 Deleting the entry ladder monitor diagram

1. To delete the entry ladder diagram, carry out the following.
   (a) Click the right-hand mouse button.
   (b) Select Delete all entry ladder (see Figure 12.6).

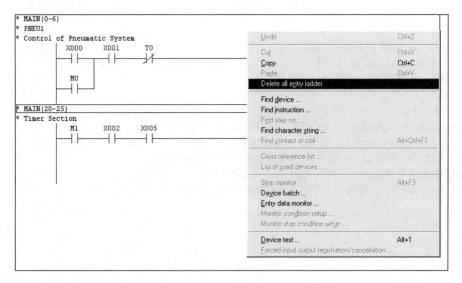

Figure 12.6

2. The entry monitor ladder diagram will now be deleted.
3. Select Main as shown in Figure 12.4 to return to the full ladder diagram.

13

Converting a MEDOC project to Gx-Developer

As MEDOC has been available for over 15 years, there will be many projects that have been produced using this software and which now, must be converted for use with Gx-Developer.

One method to convert from MEDOC to Gx-Developer is to open a new project in Gx-Developer and then upload the stored MEDOC program from the PLC.

The disadvantage with this method is that the documentation details, i.e. the comments and statements, are not usually stored in the PLC and hence the uploaded project will not contain these details.

However, if the MEDOC project is stored on disk with the ladder diagram and the documentation details, it is possible to import the complete MEDOC file directly into a Gx-Developer file.

Although comments and statements will be displayed on the imported ladder diagram, notes will not be displayed, since this function is not available with MEDOC.

13.1 Importing a MEDOC file into Gx-Developer

1. The MEDOC file to be imported will be PNEU1, which for the purpose of demonstrating this import function, will be stored in the folder C:\medoc\fxi.
2. Save and then close the Gx-Developer project currently open, i.e. PNEU1.
3. From the main toolbar, select Project.
4. Select New project.
5. Let the Gx-Developer project be named PNEU1C.
6. The New project menu is now displayed.
7. Enter the following:
    (a) PLC series   FXCPU
    (b) PLC Type   FX2N(C)
    (c) Select Ladder
    (d) Drive/path   C:\gxdevel\fxi
    (e) Project name   pneu1c.
8. Select OK.
9. Select Yes to create the project PNEU1C.
10. From the main toolbar, select Project once more and then select the following:
    (a) Import file
    (b) Import from Melsec Medoc format file.

158   *Mitsubishi FX Programmable Logic Controllers*

11. The display now becomes as shown in Figure 13.1.

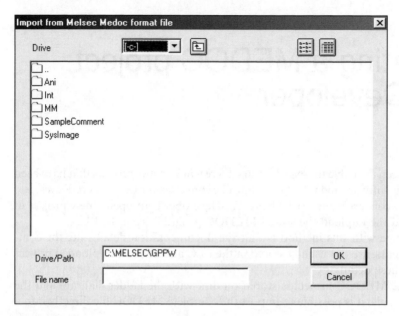

Figure 13.1

12. Repeatedly click on the yellow up arrow button shown in Figure 13.2 until the contents of the C:\root folder are displayed.

Figure 13.2

13. Select the folder C:\medoc\fxi.
14. The display now becomes as shown in Figure 13.3.
15. Select PNEU1.PRG.
16. Select OK.
17. Once the import process has been completed, the display will be as shown in Figure 13.4.
18. Select OK and the imported MEDOC project PNEU1 will be saved in Gx-Developer format as PNEU1C.
19. To include the comments and statements on the ladder diagram, select
    (a) View
    (b) Comment
    (c) Statement.
20. As mentioned earlier, notes were not part of the functionality of MEDOC and hence they will not appear on an imported MEDOC file.

Converting a MEDOC project to Gx-Developer 159

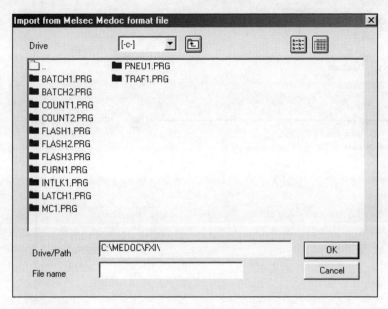

Figure 13.3

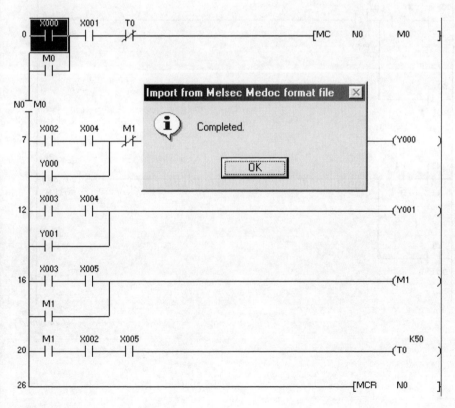

Figure 13.4

160 *Mitsubishi FX Programmable Logic Controllers*

Imported ladder diagram PNEU1C

PNEU1
Control of Pneumatic System

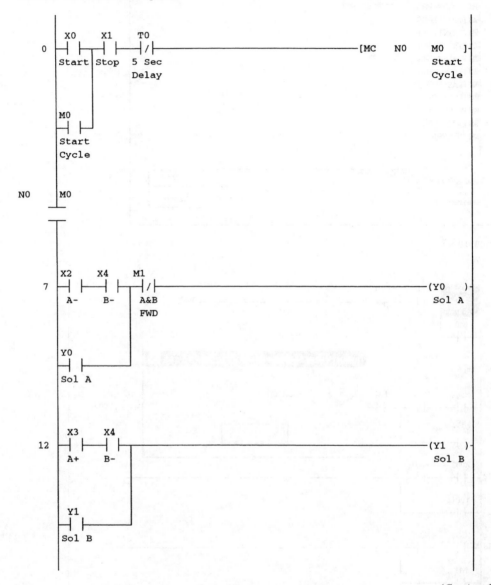

(Continued)

```
        X3    X5
  16   ─┤ ├──┤ ├──────────────────────────────────( M1    )─
        A+    B+                                    A&B
                                                    FWD


        M1
       ─┤ ├──┤
        Cycle
        Compl

*  Timer Section
        M1    X2    X5                                K50
  20   ─┤ ├──┤ ├──┤ ├──────────────────────────────( T0    )─
        Cycle A-    B+                               5 Sec
        Compl                                        Delay


  26   ──────────────────────────────────────────[ MCR  N0 ]─


  28   ──────────────────────────────────────────────[ END ]─
```

14

Change of PLC type

It is sometimes necessary to change the existing PLC type to an alternative type for example an FX1N, which is cheaper and smaller in size but has a reduced input/output specification.

The following section, describes how a project initially designed for use with an FX2N PLC, is changed to an FX1N PLC, which is a smaller and less expensive PLC.

1. Open the project PNEU1.

2. To ensure that the project PNEU1 is itself not changed, copy the project to one having a new filename, i.e. PNEU1D (see Figure 14.1).

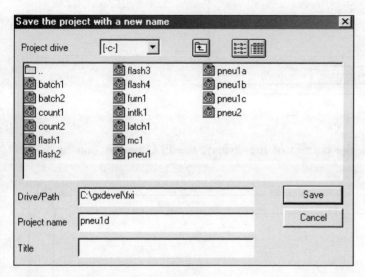

Figure 14.1

3. The project PNEU1D will now be changed to work with an FX1N PLC.
4. From the main toolbar select Project.
5. Select Change of PLC type.
6. Change the PLC type to an FX1N (see Figure 14.2).

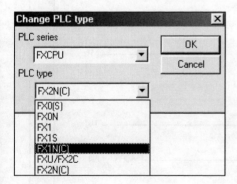

Figure 14.2

7. Select OK.
8. Select Yes, to enable the change procedure to occur (see Figure 14.3).

164 *Mitsubishi FX Programmable Logic Controllers*

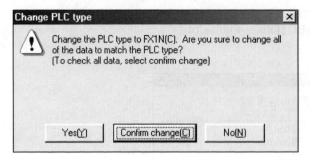

Figure 14.3

9. As can be seen from the bottom of the display, the PLC type is now an FX1N (see Figure 14.4).

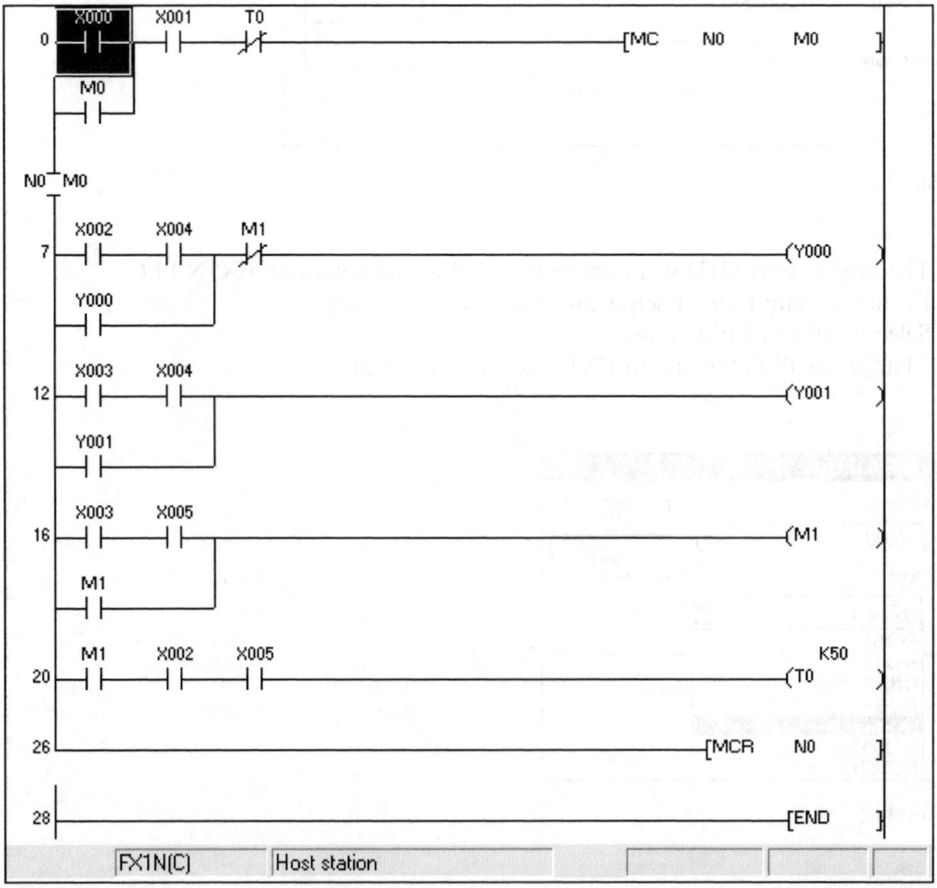

Figure 14.4

15
Diagnostic fault finding

15.1 CPU errors

Should a CPU error occur and the Error LED comes ON, then it is essential that the cause of the error be investigated and corrected.

1. Ensure a project is being displayed, i.e. PNEU1.
2. Switch the PLC to RUN.
3. From the main menu, select Diagnostics.
4. Select PLC diagnostics.
5. The display now becomes as shown in Figure 15.1.
6. The diagnostic display shows that RUN is ON.

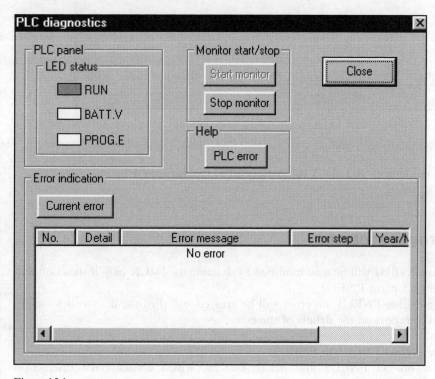

Figure 15.1

15.2 Battery error

1. Remove the FX2N CPU battery.
2. Note: On the front panel of the PLC, the battery voltage LED will now come ON.
3. The diagnostic display will now appear as shown in Figure 15.2.

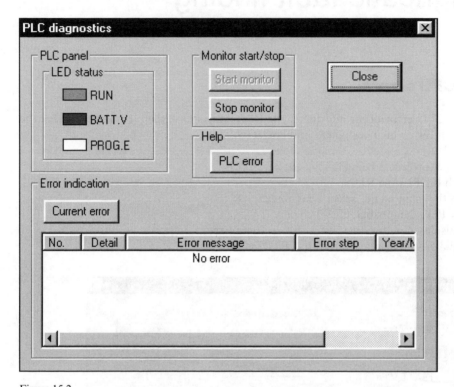

Figure 15.2

4. Replace the battery and check the battery LED has turned OFF.

15.3 Program errors

The program PNEU1 will now be modified by deleting its -[MCR N0]- instruction and given the project name PNEU3.

On downloading PNEU3, an error will be created and then the diagnostics facility will be used to report on the details of the error.

1. Copy the project PNEU1 to PNEU3.
2. Open the project PNEU3 and delete line 26, which contains the instruction -[MCR N0]-.

3. PNEU3 will now appear as shown below.

Ladder diagram – PNEU3

```
:PNEU1
:Control of Pneumatic System
```

```
          X0      X1      T0
   0     ─┤ ├────┤ ├────┤/├──────────────────────────────[MC    N0      M0   ]─
          Start   Stop    5 Sec                                          Start
                          Delay                                          Cycle
          M0
         ─┤ ├─
          Start
          Cycle

  N0      M0
  ├───────┤ ├
          Start
          Cycle

          X2      X4      M1                     * < Energise Piston A    >
   7     ─┤ ├────┤ ├────┤/├──────────────────────────────────────────────(Y0   )─
          A-      B-      A&B                                            Sol A
                          FWD
          Y0
         ─┤ ├─
          Sol A

          X3      X4                             * < Energise Piston B    >
  12     ─┤ ├────┤ ├─────────────────────────────────────────────────────(Y1   )─
          A+      B-                                                     Sol B
          Y1
         ─┤ ├─
          Sol B

          X3      X5
  16     ─┤ ├────┤ ├─────────────────────────────────────────────────────(M1   )─
          A+      B+                                                     A&B
                                                                         FWD
          M1
         ─┤ ├─
          A&B
          FWD
```

```
;Timer Section
          M1      X2      X5                                             K50
  20     ─┤ ├────┤ ├────┤ ├──────────────────────────────────────────────(T0   )─
          A&B     A-      B+                                             5 Sec
          FWD                                                            Delay

  26     ─────────────────────────────────────────────────────────────────[END ]─
```

4. Download PNEU3 and immediately the program error (PROG.E) LED will start to flash ON/OFF.
5. From the main menu, select Diagnostics.

6. Select PLC diagnostics and the display will now appear as shown in Figure 15.3.

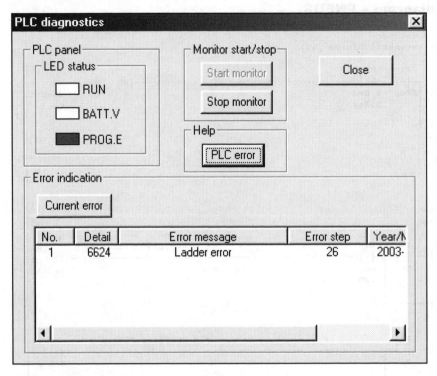

Figure 15.3

7. The diagnostic display above, shows that:
    (a) There is a ladder diagram error.
    (b) The error no. is 6624.
    (c) The error is at Step 26.
    (d) As a clock module has not been fitted to the PLC, the time and date have been taken directly from the computer.

15.4 Help display – program errors

The following section describes how a detailed description of the PLC error can be obtained.

1. Note the error number
    Error number: 6624.
2. From the PLC Diagnostics display select the PLC error box.
3. Select FX series PLC.

4. Select Open.
5. Select Error code range 6601–6632.
6. Select Open.
7. Select the Error code 6624, followed by selecting Display.
8. The error details are now shown in Figure 15.4.

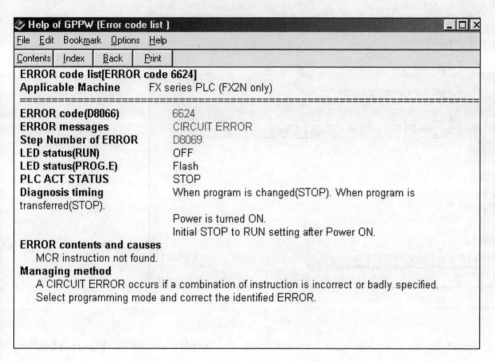

Figure 15.4

9. As can be seen from Error contents and causes, the MCR instruction MCR is missing.
10. Close down the error code list and the diagnostic display to return to the ladder diagram PNEU3.

15.5 Program error check

Since the project PNEU3 produced a program error, this can be confirmed by using the check program function in Gx-Developer as described earlier on page 28.

1. From the Tools menu, select Check program.
2. The display now becomes as shown in Figure 15.5.

Figure 15.5

3. Select Execute and as can be seen from the display shown in Figure 15.6, the instruction -[MCR N0]- at step 26, is missing.

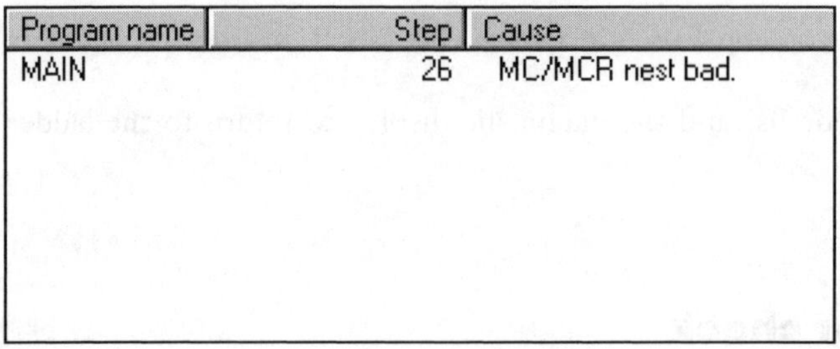

Program name	Step	Cause
MAIN	26	MC/MCR nest bad.

Figure 15.6

16

Special M coils

Within the PLC there are a group of special M coils starting at M8000.
These special coils can be used in programs to provide information such as the PLC is switched ON, timing signals, error reporting.
The list below contains just some of those which are available and which are used in this book. For a complete list, consult the Mitsubishi FX Series Programming Manual.

(a) M8000   Contacts close on switching to RUN.
(b) M8002   A one-shot pulse, whose contacts close for just one scan time, from when the PLC is switched to RUN.
(c) M8004   Error flag.
(d) M8005   Battery Low.   Internal CPU battery LOW, check battery connected. If battery is LOW, replace.
(e) M8006   Battery Low latch.
(f) M8011   0.01-sec pulses.   0.005 sec ON. 0.005 sec OFF.
(g) M8012   0.1-sec pulses.   0.05 sec ON. 0.05 sec OFF.
(h) M8013   1.0-sec pulses.   0.5 sec ON. 0.5 sec OFF.
(i) M8014   1.0-minute pulses.   30 sec ON. 30 sec OFF.
(j) M8034   Disable all Y outputs.
(k) M8022   Carry   used with arithmetic and shift instructions.
(l) M8200–M8234   used with bi-directional counters.
(m) M8235–M8255   used with high-speed counters.

16.1 Device batch monitoring

Device batch monitoring, which enables a block or batch of devices to be monitored, will now be used to monitor the special M coils.

1. Open a project, i.e. C:\gxdevel\fxi\count1 (Refer page 82).
2. Select the following:
   (a) Online
   (b) Monitor
   (c) Device batch.
3. The display now becomes as shown in Figure 16.1.
4. Enter M8000.
5. Select Start monitor.

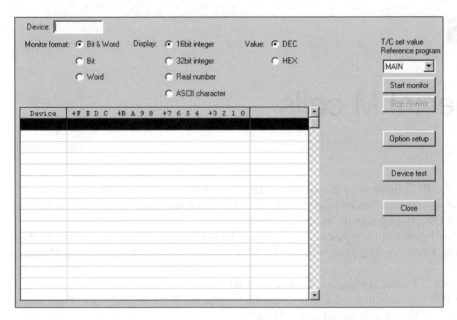

Figure 16.1

6. Switch the PLC to RUN.
7. The monitored display now of all the M coils from M8000 to M8159 are shown in Figure 16.2.

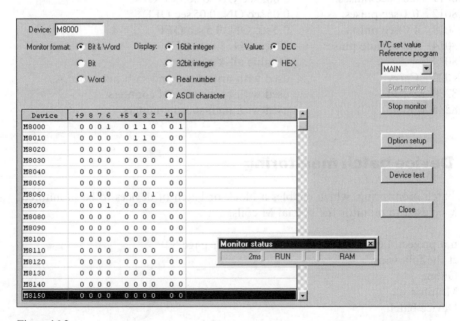

Figure 16.2

8. Check the logic state of the above special M coils with the list on page 171.

Special M coils 173

16.2 Option setup

The option setup enables the following to be selected:

1. The reversal of the bit order display.
2. The number of points can be either 10 or 16.
3. Select the Options shown in Figure 16.3.

Figure 16.3

4. The display now becomes as shown in Figure 16.4, with as can be seen, the logic state of M8000 displayed in the top left-hand corner of the display.

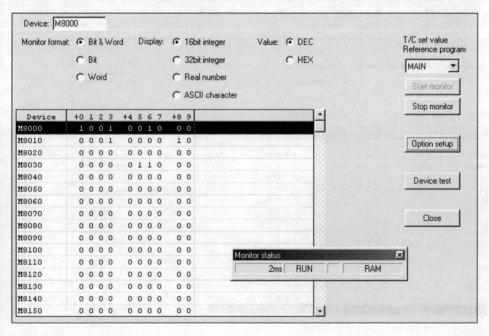

Figure 16.4

16.3 Monitoring the X inputs

The X inputs will now be monitored again using device batch.

1. Operate the inputs X1, X2, X4.
2. Enter 0099 on the thumbwheel switches. These are connected to X10 to X27.
3. Enter X0 into the device window.
4. Configure the Option setup (see Figure 16.5) as follows:
   (a) Bit order – F-0
   (b) Display – 16 bit integer.

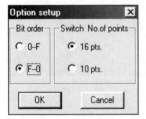

Figure 16.5

5. Configure the Monitor display as:
   (a) Monitor format – Bit 8 Word
   (b) Value – HEX (This is the hexadecimal equivalent of the binary pattern X0 to X17).
6. The display now becomes as shown in Figure 16.6.

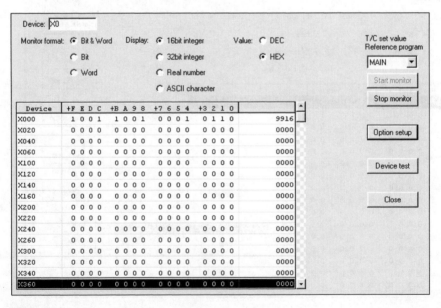

Figure 16.6

Note: The device header for the binary pattern as shown in Figure 16.6 is incorrect, i.e. X0 to XF. This is because the inputs for the FX range of PLCs are in octal, i.e. X0 to X17.

17

Set–reset programming

Set–reset programming is an extension to conventional ladder programming, which enables the easier programming of continuous, sequential PLC control systems.

The complete process is broken down into a number of discrete steps. The output from each step being, either a Set or a Reset instruction.

17.1 PNEU4

The program PNEU4 carries out the same sequence of operations as PNEU1, but unlike PNEU1, which uses a standard ladder diagram, PNEU4 uses set–reset programming techniques.

Basic system

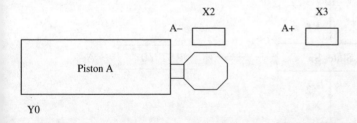

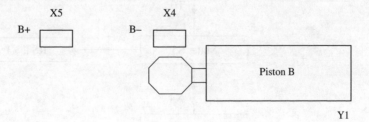

Figure 17.1

17.2 Sequence of operation – automatic cycle

The sequence of operations for the two pistons is as follows:

1. Press   Start.
2. A+   Piston A OUT.
3. B+   Piston B OUT.
4. A−   Piston A IN.
5. 5-sec time delay.
6. B−   Piston B IN.
7. Repeat sequence.

17.3 Sequence function chart – PNEU4

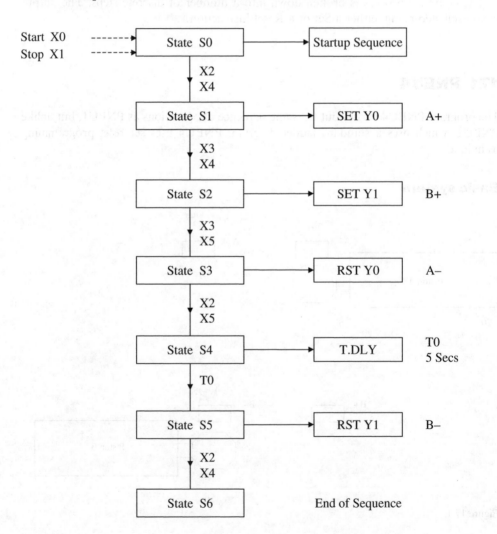

17.4 Ladder diagram – PNEU4

From the sequential function chart, the following set–reset ladder diagram can be produced.

```
      X0      X1      T0
 0    ┤ ├────┤ ├────┤/├─────────────────────────────( M0    )
      Start  Stop
      M0
      ┤ ├

      M0      X2      X4                      *<Sol A ON
 5    ┤ ├────┤ ├────┤ ├──────────────────────[SET    Y0   ]─A+
              A-      B-

      M0      X3      X4                      *<SOL B ON
 9    ┤ ├────┤ ├────┤ ├──────────────────────[SET    Y1   ]─B+
              A+      B-

      X3      X5                              *<SOL A OFF
13    ┤ ├────┤ ├─────────────────────────────[RST    Y0   ]─A-
      A+      B+
      M0
      ┤/├

      X2      X5                                            K50
17    ┤ ├────┤ ├──────────────────────────────────────(T0       )

      T0                                      *<SOL B OFF
22    ┤ ├───────────────────────────────────────[RST    Y1   ]─B-
      M0
      ┤/├

25    ────────────────────────────────────────────────────[END ]─
```

17.5 Principle of operation

1. Line 0
   (a) Inputs X0 and X1 are from the start and cycle stop buttons respectively.
   (b) The operation of input X0 will cause the internal relay M0 to energise and latch over its own contact.
   (c) The operation of the stop input or the timer contact T0 will break the latch circuit.
2. Line 5
   (a) With the two Pistons in the A– and B– positions, inputs X2 and X4 will be CLOSED. Therefore as soon as M0 closes, output Y0 will be SET.

178 Mitsubishi FX Programmable Logic Controllers

 (b) The use of the instruction SET enables an output coil to be turned ON and to remain ON, even when the input condition is no longer present.
 (c) Therefore when piston A moves forward and input X2 opens, output Y0 will still remain energised.
3. Line 9
 (a) When piston A reaches the A+ position, then input X3 will CLOSE.
 (b) Since piston B is in the B− position, input X4 will be CLOSED and therefore output Y1 will now be SET and piston B will also be energised.
 (c) Hence at this moment in the cycle, both pistons A and B will be energised.
4. Line 13
 (a) When piston B reaches the B+ position, input X4 will be OPEN and inputs X3 and X5 will be CLOSED.
 (b) This will RESET output Y0 and hence piston A will return to the A− position.
 (c) The function of the normally closed M0 contacts is to ensure that Y0 will be RESET, when the STOP is operated or when the PLC is first turned ON.
5. Line 17
 (a) With the piston A returning to its A− position, input X3 will be OPEN and input X2 will reclose, this plus input X5 will cause timer coil T0 to be energised.
 (b) This will cause timer T0 to start timing out.
6. Line 22
 (a) When timer T0 times out after 5 sec, its T0 contact will CLOSE and this will RESET output Y1.
 (b) Piston B, will now return to the B− position.
 (c) Input X5 will now open and this will cause timer T0 to RESET.
 (d) The function of the normally closed M0 contacts is to ensure that Y1 will be RESET, when the STOP is operated or when the PLC is first turned ON.
7. Line 0
 (a) During the short time that timer T0 is energised, i.e. whilst piston B is returning to the B− position, its normally closed contact will be OPEN. This will cause the M0 latch circuit to drop out.
 (b) The system will now HALT, until the START button is reoperated.

17.6 Simulation and monitoring procedure

1. **Display and monitor the Ladder Diagram PNEU4.**
2. **Operate the Input Switches X1, X2 and X4.**
 This will simulate the operation of the Stop push button and the A− and B− limit switches.
3. **Momentarily operate the Start Switch X0.**
 Output Y0 wil now be energised and this will cause Piston A to operate to the A+ position.
 In a real situation with Piston A moving forward, its limit switches X2 would open and X3 would close.

4. **Open X2 and Close X3.**
   This will cause Output Y1 to energies and hence enable Piston B to move forward to its B+ position.
5. **Open X4 close X5.**
   With Inputs X3 and X5 now closed output Y0 is reset.
6. **Open X3 and close X2.**
   With X2 closing, this will start the operation of Timer T0.
   After 5 seconds, Timer T0 will time out and Reset Y1. Hence Piston B will return to its B− position.
7. **Open X5 and Close X4.**
   The process can now be repeated, by pressing the Start push button.

17.7 Monitoring PNEU 4

Using entry data monitor, monitor the elements shown in Figure 17.2.

Device	ON/OFF/Current	Setting value	Connect	Coil	Device comment
X000			0		Start
X001			1		Stop
M0			0		
X002			1		A−
X003			0		A+
X004			1		B−
X005			0		B+
Y000			0		
Y001			0		

Figure 17.2

18

Trace

Trace is a method by which historical trend graphs can be produced.

The displays can show graphically the operations of selected input/output devices and also the values stored in timers, counters and data registers.

It is used in fault finding rapidly changing signals, which cannot be observed with input/output LEDs and can be used to display the logic conditions of input/output devices and the contents of timers, counters and registers before and after the occurrence of the fault condition.

To demonstrate the Trace facility, project PNEU4 will be used.

18.1 Principle of operation

Block diagram – Multi-channel shift register

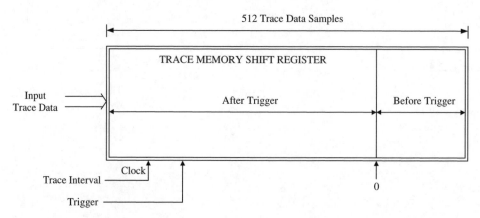

Figure 18.1

The basis of Trace is that RAM memory within the PLC is allocated for use as a multi-channel shift register (see Figure 18.1).

The input Trace data can consist of up to 10-bit devices, i.e. X, Y. M and up to 3×16-bit word devices, i.e. timers, counters and data registers.

Each time a clock signal is applied to the Trace memory shift register, a sample of data, i.e. the logic state of the bit devices and the contents of the word devices within the PLC, is clocked into the register.

Each sample of stored information is known as a Trace.

With no trigger signal, the input Trace data is clocked into the Trace memory shift register and when it is full with 512 Traces, the data is clocked out and is lost.

When a trigger signal is applied, only the required number of Traces after the trigger signal, will now be clocked into the shift register. Hence stored in the Trace memory will be a total of 512 Traces, which is a record of the events prior to and after the Trigger signal.

When all of the 'After Trigger Traces' have been clocked in, then stored in the shift register will be a number of Traces, which occurred before the Trigger signal, i.e. 64, and a selected number of Traces, i.e. 448, which occurred after the Trigger signal.

The trigger signal can be obtained for example, from a PLC input, when a fault condition occurs.

The stored data in the Trace memory shift register is then uploaded into Gx-Developer and can be displayed in graphical format before and after the fault condition.

Note: It will be shown later in the Trace conditions section how the above trace numbers are obtained.

18.2 Ladder diagram – PNEU4

```
         X0       X1       T0
  0      ─┤├──────┤├──────┤/├──────────────────────────────────────(M0    )─
         Start   Stop
         M0
         ─┤├─

         M0       X2       X4                          *<Sol A ON
  5      ─┤├──────┤├──────┤├──────────────────────────────[SET    Y0  ]─
                  A-       B-

         M0       X3       X4                          *<SOL B ON
  9      ─┤├──────┤├──────┤├──────────────────────────────[SET    Y1  ]─
                  A+       B-

         X3       X5                                   *<SOL A OFF
 13      ─┤├──────┤├──────────────────────────────────────[RST    Y0  ]─
         A+       B+
         M0
         ─┤/├─

         X2       X5                                                K50
 17      ─┤├──────┤├──────────────────────────────────────────(T0       )─

         T0                                            *<SOL B OFF
 22      ─┤├──────────────────────────────────────────────[RST    Y1  ]─
         M0
         ─┤/├─

 25      ─────────────────────────────────────────────────────[END ]─
```

18.3 Trace setup procedure

To obtain a Trace it is necessary to carry out the following procedure:

1. From the online menu, select:
   (a) Trace
   (b) Sampling Trace.
2. The display now becomes as shown in Figure 18.2.

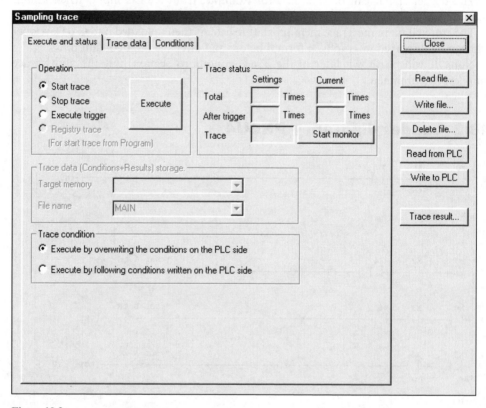

Figure 18.2

18.4 Trace data

1. Select Trace data.
2. The Trace data are those devices, which will be monitored and displayed as Trace waveforms.
3. Enter the following devices into the Trace data menu as shown in Figure 18.3.
   (a) X0
   (b) X2
   (c) X3

(d) X4
(e) X5
(f) Y0
(g) Y1
(h) T0

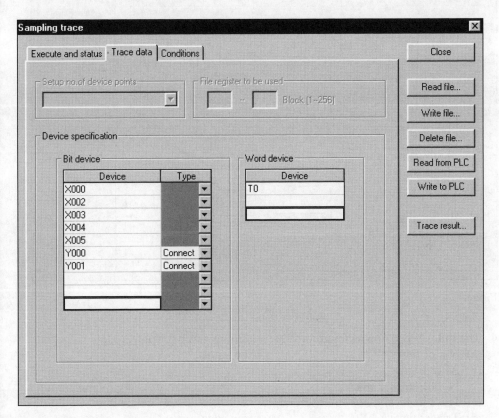

Figure 18.3

18.5 Trace conditions

1. Select Conditions.
2. The display now appears as shown in Figure 18.4.
3. The conditions, which can be set, are as follows:
   (a) The total number of times (Traces), in which the state or value of the selected devices is monitored and saved to the Trace memory.
   (b) The number of Traces that are to be recorded after the trigger signal has occurred.
   (c) The trigger signal that will be used to start the Trace. This will be the input X0.
   (d) The time period, for repetitive monitoring of the devices.

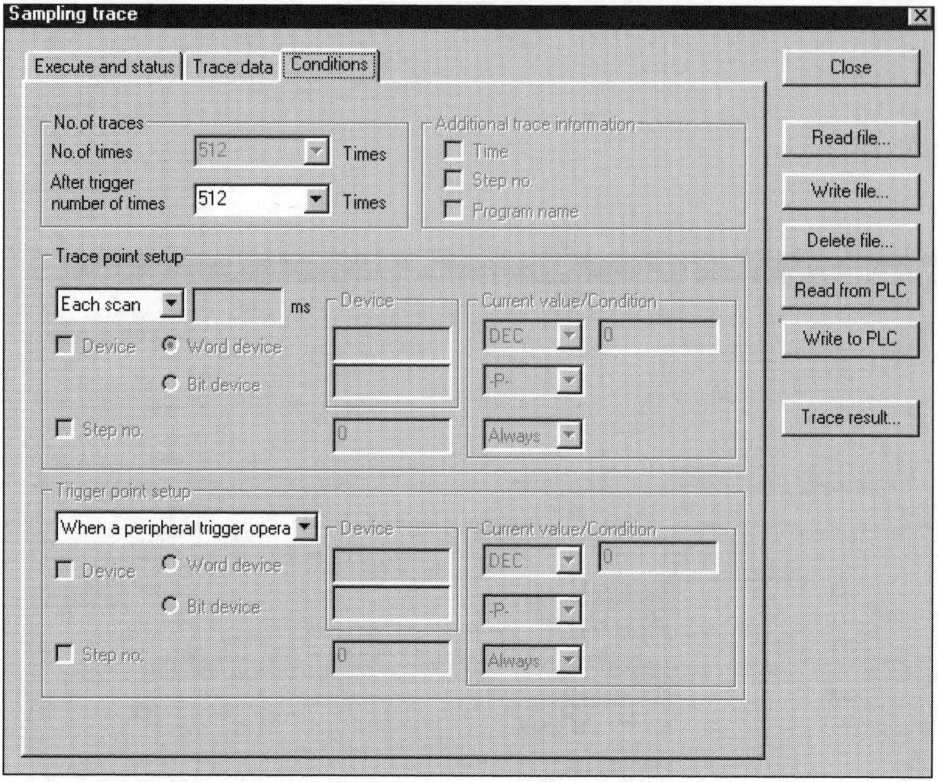

Figure 18.4

4. The following is now entered into the conditions setup menu.
   (a) No. of times   512   fixed value
   (b) After trigger
       No. of times   448
   (c) Trace interval 100 milli-seconds
   (d) Trigger device X0   rising edge.
5. This means:
   (a) The total Trace time will take 512 × 100 milli-seconds = 51.2 sec.
   (b) The number of Traces before the trigger signal (X0) will be 64, i.e. (512−448).
   (c) The number of Traces after the trigger signal will be 448.
6. This ensures that the waveforms can be examined for a short time before the trigger signal occurs, i.e. the closure of input X0.
7. After the condition data has been entered, the conditions display will appear as shown in Figure 18.5.

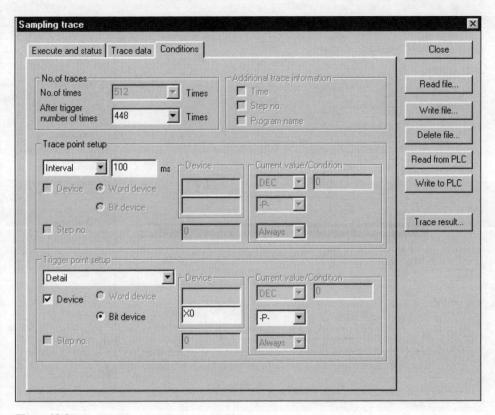

Figure 18.5

18.6 Transfer Trace data to PLC

1. Ensure the sampling Trace window is being displayed.
2. Select Write to PLC, to download the Trace data and the Conditions into the PLC's memory.
3. A section of the PLC's RAM memory is now configured for storing the Trace data.

18.7 Saving the Trace setup data

The Trace setup data can now be saved on the hard disk of the computer and recalled when required.

1. From any of the Trace menus, select Write file.
2. The display now becomes as shown in Figure 18.6.
3. Enter the file name PNEU4.
4. Select OK.

186 *Mitsubishi FX Programmable Logic Controllers*

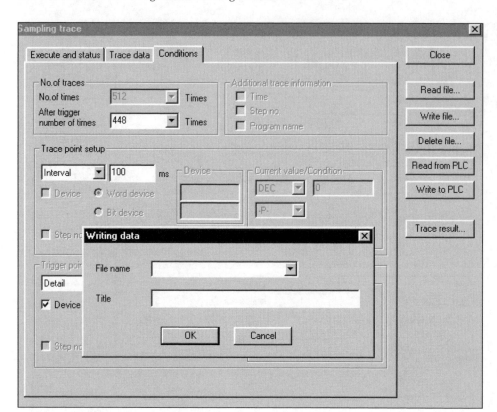

Figure 18.6

5. Select OK to return to the Trace data menu.
6. The Trace setup data has now been saved to the Trace file PNEU4.

18.8 Reading the Trace setup data from file

The Trace setup data, which was previously written to the Trace file PNEU4, will now be recalled by reading the file.

1. Select Read file (see Figure 18.7).
2. Click on the down arrow button ▼ in the Reading data window.
3. Select PNEU4.
4. Select OK.
5. As soon as the file PNEU4 has been read into the project, select OK to return to the Trace menus.
6. Note that the Trace setup data is now displayed in all of the menus.

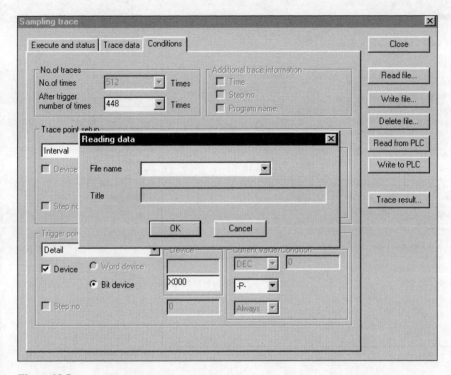

Figure 18.7

18.9 Start Trace operation

The following procedure is now used to obtain the Trace display:

1. The very first time Trace is operated, it is necessary to execute the trigger condition.
2. (a) Select the Execute and status menu
   (b) Select Execute trigger.
3. Select Execute.
4. The display now becomes as shown in Figure 18.8.
5. Select the Start Trace button from the Execute and status menu.
6. Select Execute to start the Trace process.
7. A few seconds later the display appears, with the number of Traces stored. This means that the changing logic conditions of the following devices are being stored as Traces within the FX2N PLC memory, prior to the start trigger signal.
   (a) Inputs X0, X2, X3, X4, X5
   (b) Outputs Y0 and Y1
   (c) The contents of timer T0.
8. Until the trigger condition, i.e. X0 occurs, the memory will store only the last 512 Traces of data, i.e. the last 51.2 sec of data. As more Traces fill up the memory, then the earlier Traces will be lost.
9. Wait until the Trace memory has filled up with 512 Traces, this will take 51.2 sec (see Figure 18.9).

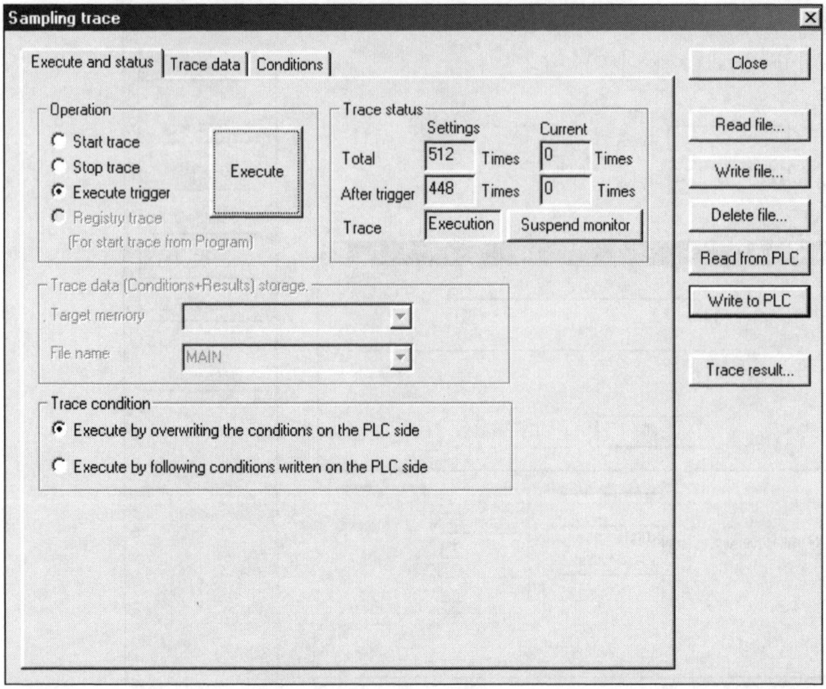

Figure 18.8

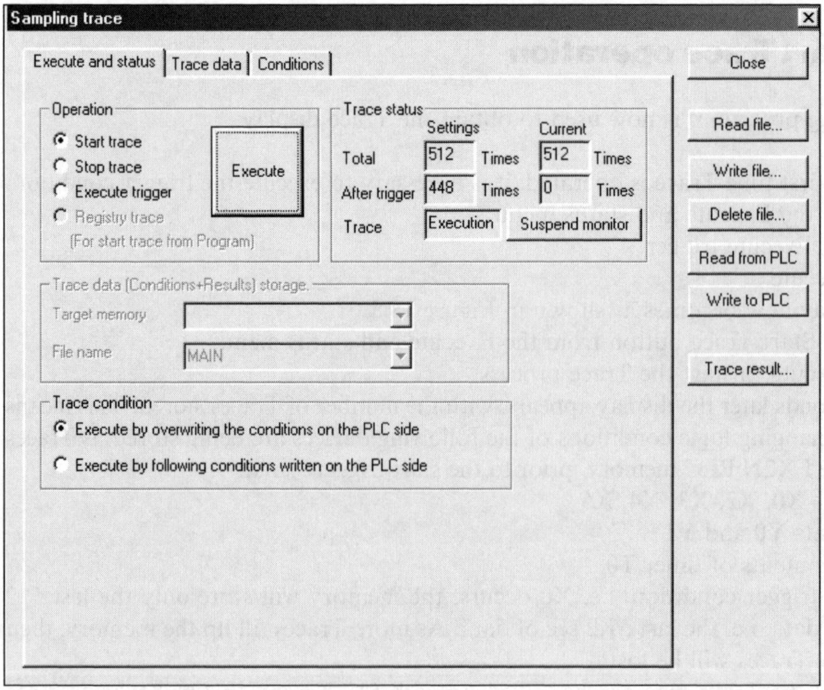

Figure 18.9

18.10 Start trigger – X0

1. Operate the input switches X1, X2 and X4.
2. Momentarily operate the input X0, which will cause the following to occur:
   (a) Start the pneumatic operation by turning Y0 ON.
   (b) Provide the Trace trigger signal.
3. Note the number of Traces, which have been stored in the Trace memory after the occurrence of the trigger signal X0. This is shown in the box 'Current Traces after Trigger', i.e. 11 (see Figure 18.10).

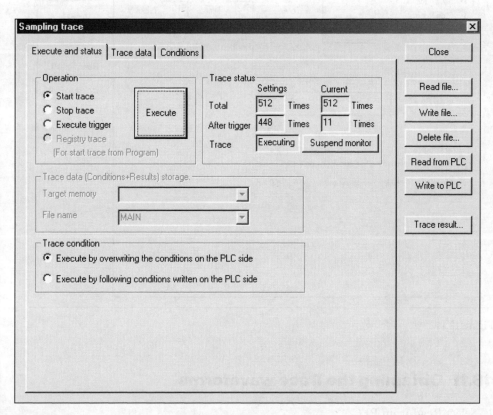

Figure 18.10

4. Carry out the remainder of the pneumatic operation.
   (a) Open X2 and close X3 – this will cause Y1 to energise.
   (b) Open X4 and close X5 – this will cause Y0 to de-energise.
   (c) Open X3 and close X2 – this will start the operation of timer T0.
   (d) After 5 sec, timer T0 times out – this will cause Y1 to de-energise.
   (e) Open X5 and close X4 to complete the machine cycle.
5. After the operation of the Trace signal X0, a further 448 Traces will be clocked into the Trace memory shift register to complete the Trace process.

6. A total of 512 Traces will now be stored into the Trace memory, i.e. the 64 Traces prior to the Trace signal X0 and the 448 Traces after the operation of the trigger signal (see Figure 18.11).

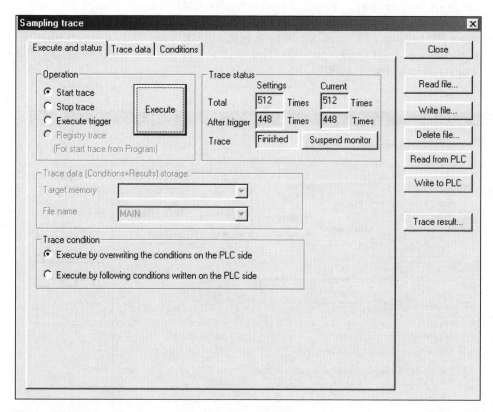

Figure 18.11

18.11 Obtaining the Trace waveforms

1. Select Read from PLC.
2. When the read of the Trace information from the PLC has been completed, select OK.
3. Select Trace result, to obtain the Trace displays.

18.12 Trace results

1. The Trace display initially appears, as shown in Figure 18.12.
2. Change the Display units to 100.

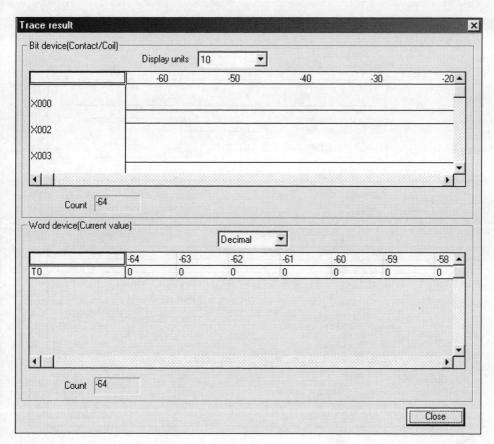

Figure 18.12

3. The display should now appear as shown in Figure 18.13.
4. Move the vertical cursor downwards, so that the other Trace waveforms can be viewed.
5. The display shown in Figure 18.14 is a combination of the Trace waveforms.
6. From an examination of the waveforms, the operation of PNEU4 can be determined.
7. Using the Trace waveforms in Figure 18.14, confirm the following sequence of events.
   (a) With the operation of X0, Y0 turns ON.
   (b) X2 then opens and X3 closes.
   (c) With X3 closing, Y1 turns ON.
   (d) X4 then opens and X5 closes.
   (e) With the closure of X5, Y0 turns OFF.
   (f) X3 then opens and X2 closes.
   (g) There is now a 5-sec delay until Y1 turns OFF.
   (h) X5 then opens and X4 closes.

192   *Mitsubishi FX Programmable Logic Controllers*

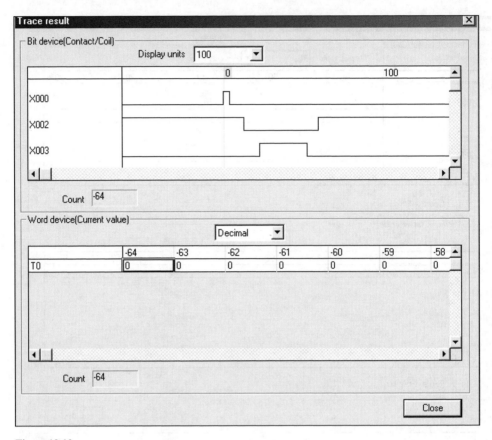

Figure 18.13

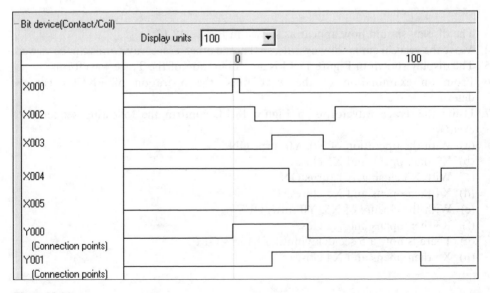

Figure 18.14

18.13 Measuring the time delay – T0

The time measurements between any parts of the machine cycle can now be obtained, from the Trace waveforms.

The following procedure describes how the 5-sec time delay of T0, can be checked from the Trace waveforms.

1. The 5-sec delay of T0 is controlled as follows:
   (a) It starts with the closure of X2, i.e. when piston A returns to its A– position.
   (b) On timing out, output Y1 turns OFF.
2. Using the left-hand mouse button, double click on the rising edge of X2 (see Figure 18.15).

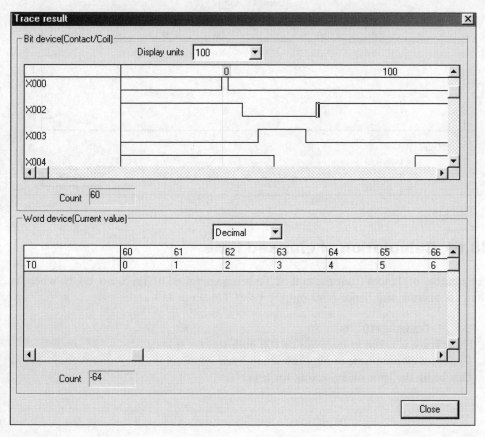

Figure 18.15

3. Note that timer T0 has been energised at Trace no. 60.
4. Now double click on the falling edge of Y1 (see Figure 18.16).
5. Note that timer T0 times out at Trace no. 110.

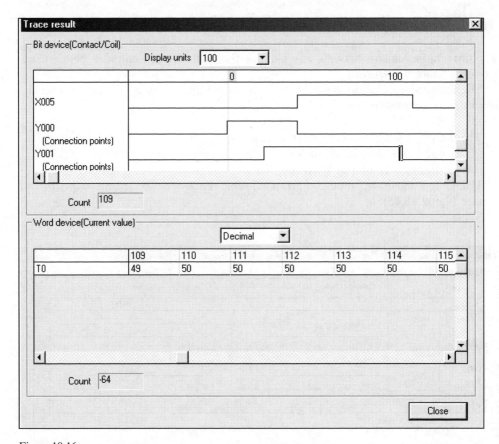

Figure 18.16

18.14 Calculation of elapsed time

The number of Traces from the coil of T0 being energised at Trace no. 60, to when its contacts operate and hence reset output Y1 at Trace no. 110, is

1. No. of Traces = 110 − 60 = 50
2. Each Trace is taken at intervals of 100 milli-sec, i.e. 0.1 sec.
3. Therefore elapsed time = 50 × 0.1 sec = 5 sec
4. This being the time delay setting for timer T0.

19

Data registers

The PLC programs produced so far, basically replace conventional electromagnetic relay systems plus additional timers and counters, which are normally separate plug-in units.

However, the programs that will now be produced are far more complex and enable the PLC to become more identified with computer type operations.

The starting point for these programs is the data register.

The data register is equivalent to 16 auxiliary memory coils, which are linked together to store in binary form, numerical values.

In addition it is possible to link two data registers together, i.e. the equivalent of 32 M coils, to enable even larger numbers to be stored.

Register size	Number range
Single – 16 bits	−32767 to +32768
Double – 32 bits	−2147 million to +2147 million

Format – 16 bit data register

b15 ──────────────── b0

b0   least significant bit
b15 most significant bit

Each bit of the data register stores either a logic 0 or a logic 1.

19.1 Number representation – binary/decimal

In the binary system only the values 0 and 1 can be used, but by using a number of binary bits it is possible for them to be equal to any standard denary (decimal) number.

Decimal value	Binary value
0	0000 0000 0000 0000
1	0000 0000 0000 0001
2	0000 0000 0000 0010
3	0000 0000 0000 0011
4	0000 0000 0000 0100
5	0000 0000 0000 0101
6	0000 0000 0000 0110
7	0000 0000 0000 0111
8	0000 0000 0000 1000
9	0000 0000 0000 1001
10	0000 0000 0000 1010
⋮	⋮
255	0000 0000 1111 1111
256	0000 0001 0000 0000
257	0000 0001 0000 0001
⋮	⋮
4095	0000 1111 1111 1111
4096	0001 0000 0000 0000
4097	0001 0000 0000 0001
⋮	⋮
32767	0111 1111 1111 1111
−1	1111 1111 1111 1111
−32768	1000 0000 0000 0000

Note: Where the most significant bit is a logic 1, then this signifies a negative number.

19.2 Converting a binary number to its decimal equivalent

It is possible to convert a binary number to its denary equivalent, by assigning a denary weighting value to each of the binary columns. For ease of understanding, only 8-bit numbers will be considered.

The denary weighting values are as follows:

Column number	Weighting value
1	1
2	2
3	4
4	8
5	16
6	32
7	64
8	128

Example: Convert the binary number 1001 1101 to its denary equivalent.
Note: msb is the Most Significant Bit. lsb is the Least Significant Bit.

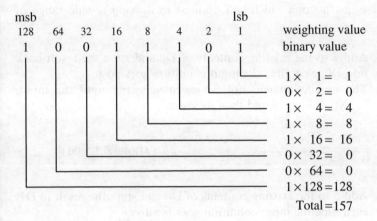

Hence 1001 1101 = 157

19.3 Binary numbers and binary coded decimal

A binary value can be displayed in two different ways:

(a) Pure binary
(b) Binary coded decimal (BCD).

Pure binary

Consider the following binary number 0010 0101 1001 0110.
   In pure binary 0010 0101 1001 0110 = 9622

Binary coded decimal

In BCD the binary numbers are grouped into blocks of four binary digits each starting from the lsb. Each block of binary digits is then converted to its denary equivalent.

```
1001   0110   0010   0010
  |      |      |      |
  9      6      2      2
```

   In BCD, the binary pattern would now have to be 1001 0110 0010 0010 for it to equal 9622. The reason for using BCD is that each block of four digits can be output to a binary to decimal decoder chip, whose outputs can then be connected to a four digit 0–9999 display.

19.4 Advanced programming instructions

The advanced instructions, which are briefly described in this section all use data registers, and it is these instructions which will be used to develop a wide range of PLC programs.

1. **ADDP D0 K2 D0**   Add 2 to the existing contents of D0 and store the result back into D0, each time the input condition goes positive.
   The instruction will not be executed again, until the input, i.e. X0, is opened and then reclosed.

   ```
        X0
   |----| |----------------------------------------------[ ADDP  D0  K2  D0 ]|
   ```

2. **ADDP D0 K5 D1**   Add 5 to the existing contents of D0 and store the result in D1, each time the input condition goes positive.
3. **BCD D0 K4Y10**   Transfer 16 bits of data from D0 to Y10–Y27 in BCD format, i.e. from 0–9999, each time the instruction is executed. This instruction would be used to display the contents of data register D0, on a four-digit seven-segment display unit, which had been connected to outputs Y10–Y27.

   ```
        X0
   |----| |----------------------------------------------[ BCD  D0  K4Y10 ]|
   ```

   While the input, i.e. X0, is closed, the instruction will be executed during each scan of the program, i.e. every few milli-seconds.
   Where a K value is linked to another element, then the k value indicates the number of bits to be used with that element.
   k1 4 bits
   k2 8 bits
   k3 12 bits
   k4 16 bits

4. **BIN K4X10 D0**   Load data register D0 with 16 bits of data from X10–X17 in BCD format.
   bin = BCD INPUT Input number range 0–9999.
   This instruction would be used to input numerical values from a bank of four-decimal 0–9 thumbwheel switches, connected to X10–X27.
5. **CMP D0 K10 M2**   Each time this instruction is executed. The logic states of the internal memory coils M2, M3 and M4, will indicate the result of the comparison.
   M2 ON if D0 > 10
   M3 ON if D0 = 10
   M4 ON if D0 < 10

Note: This instruction is the only type of compare instruction available, when using the FX range of PLCs. However, when using the FX1S, FX1N and the FX2N range of PLCs the inline comparison instructions can be used, hence saving on M Coils.

6. Inline comparison instructions – see Note above. If the comparison is TRUE, then continue to execute the remaining instructions on that rung including the output. If the comparison is FALSE, then move to execute the instructions on the following rung.
   (a) -[> D0 K10]- Continue if the contents of D0 are greater than 10.
   (b) -[>= D0 K10]- Continue if the contents of D0 are greater than or equal to 10.
   (c) -[> D0 K10]- Continue if the contents of D0 are equal to 10.
   (d) -[< D0 K10]- Continue if the contents of D0 are less than 10.
   (e) -[<= D0 K10]- Continue if the contents of D0 are less than or equal to 10.

7. **DECP D0** — Decrement the contents of D0, each time the input condition goes positive.

8. **DIVP D16 K120 D18** — Divide the existing contents of D16 by 120 and store the result in D18 and D19, each time the input condition goes positive. Quotient part of the result stored in D18. Remainder part of the result stored in D19.
   For example if D16 contains 345, then
   D18 (quotient) = 2
   D19 (remainder) = 105

9. **FMOV K0 D0 K4** — Clear four data registers from D0 to D3 inclusive.

10. **INCP D0** — Increment the contents of data register D0, each time the input condition goes positive.

11. **MOV D0 K4Y0** — Transfer 16 bits of data from D0 to Y0–Y17, in binary format, i.e. from 0–65535.

12. **MOV K0 D0** — Move, i.e. load the value 0 into data register D0.

13. **MOV K104 D0** — Load the value 104 into data register D0.

14. **MOVP K0 D0** — Clear data register D0, each time the input condition goes positive.

15. **MOV K2X0 D0** — Load data register D0 with 2 × 4 bits of data from X0 to X7. Input number range 0–255.

16. **MULP D2 K10 D3** — Multiply the existing contents of D2 by 10, each time the input condition goes positive.
    Lower 16 bits of the result are stored in D3.
    Upper 16 bits of the result are stored in D4.

17. **RST D0** — Clear the contents of data register D0. Identical to -[MOV K0 D0]-.

18. **SUBP D0 K5 D1** — Subtract 5 from the existing contents of D0 and store the result in D1, each time the input condition goes positive. D1 = D0 − 5.

19. **-(T0 ^{D0})-** — Load timer T0 with the contents of data register D0. This is used to change the value of the time delay, whilst the program is being executed.

20

Introduction to programs using data registers

The programs which will be developed in this section are designed to enable the user to gain experience and hence confidence in using a few of the more easily understandable advanced programming instructions.

20.1 Binary counter – COUNT3

A simple counting system will now be produced, in which a push button will simulate pulses coming from a transducer.

In an industrial application, the transducer could be monitoring the rotation of a toothed gear wheel connected to, for example, a conveyor system.

The total number of pulses is displayed on the output LEDs in pure binary form.

System – block diagram

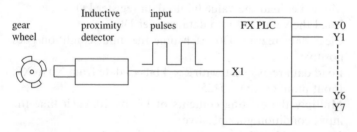

Figure 20.1

As the toothed wheel rotates, the difference in the width of the gap between the teeth and the inductive proximity detector will change. Within the detector is a small magnet and coil and as the air gap between the gear wheel and the detector changes, it will cause a change in the magnetic field within the detector, which in turn, will induce voltage pulses into the coil (see Figure 20.1).

These pulses are now fed into the PLC, where they can be:

(a) Counted.
(b) Displayed as an incrementing binary value on the Y Outputs, Y0–Y7.

Ladder diagram – COUNT3

This counter circuit will be used to count the incoming pulses and display the incrementing count in pure binary form on the Output LEDs Y0 to Y7.

Principle of operation

1. Line 0
   The closing of input contact X0 will cause the data register D0 to be cleared by loading it with 0. The use of 'P' ensures that this operation only occurs when the input X0 is operated and not each time the program is scanned.
2. Line 6
   The opening and closing of input X1 simulates the input pulses to the PLC. Each time input X1 is closed, the contents of data register D0 are incremented by one.
3. Line 10
   Input M8000 is closed while the PLC in RUN and therefore during each scan of the program, the changing contents of D0 will be transferred to the eight outputs Y0 to Y7.

Entering the program

Most advanced instruction can be entered, simply by typing them in from the keyboard. Hence, the program COUNT3 can be entered as shown below.

Instruction		Keyboard entry	
1. Normally open contact X0	1	x0	<ent>
2. MOVP K0 D0		movp	<space>
		k0	<space>
		d0	<ent>
3. Normally open contact X1	1	x1	<ent>
4. INCP D0		incp	<space>
		d0	<ent>
5. Normally open contact M8000	1	m8000	<ent>
6. MOV D0 K2Y0		mov	<space>
		d0	<space>
		k2y0	<ent>
7. END		end	<ent>

202 *Mitsubishi FX Programmable Logic Controllers*

Program monitoring

To monitor the operation of the program in real time, carry out the following operations:

1. Download the program COUNT3 to the FX2N PLC.
2. Display the ladder diagram.
3. Press F3 for monitor mode.
4. The display now becomes as shown in Figure 20.2, after six input pulses.

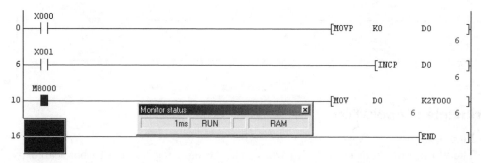

Figure 20.2

5. The ladder diagram display shows that in addition to the inputs and outputs being monitored, so can the contents of data register D0 be monitored.

20.2 BCD counter – COUNT4

Each time the input X1 is operated, the 4 × 7 segment display connected to Y10–Y27 will show an incrementing count in BCD format.

That is the output number can only be in the range 0–9999.

Principle of operation – COUNT4

1. Line 0
   Each time input X0 is operated, data register D0 is cleared to zero.
2. Line 6
   Each time input X1 is operated, the contents of data register will be incremented, i.e. increased by one.
3. Line 10
   While input X0 remains operated, the 16-bit contents of D0 are transferred, i.e. copied in BCD form to the 16 Y outputs, Y10–Y27.

Hence displayed on the seven-segment display will be an incrementing count in BCD format from 0 to 9999.

Monitoring

1. Press F3 and monitor the ladder diagram.
2. Enter 16 input pulses to obtain the display show below.

3. Using device batch, monitor the outputs Y10 to Y27 (see Figure 20.3).

BCD display

1. The binary display is in BCD format, which can easily be converted into the decimal equivalent of 16. However the display shows that the decimal equivalent is 22, which is incorrect. This is because it is the straight decimal equivalent of the displayed binary pattern.
2. To obtain the correct decimal equivalent of a BCD value, select the HEX button.
3. As shown in Figure 20.4, the decimal equivalent of the binary pattern is now the correct value, i.e. 16.

204 *Mitsubishi FX Programmable Logic Controllers*

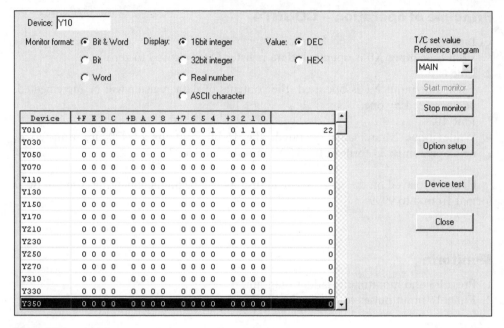

Figure 20.3

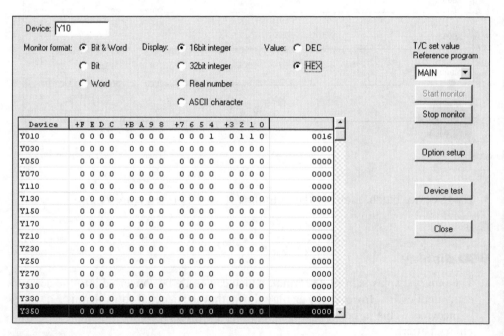

Figure 20.4

Introduction to programs using data registers 205

20.3 Multiplication program – MATHS 1

The project MATHS 1 will enable two numbers to be multiplied together and then display the result on the seven-segment displays.

1. Enter from the thumbwheel switches two separate numbers in the range 0–99.
2. Multiply the two numbers together.
3. Display the result, which will be in the range 0–9801, on the seven-segment displays.

Ladder diagram – MATHS 1

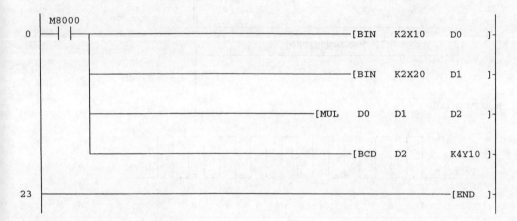

Principle of operation

1. The four-digit number from the thumbwheel switch, i.e. 9876 is split into two separate numbers.
2. N1 = 76 and is stored into D0.
   N2 = 98 and is stored into D1.
3. The two numbers are then multiplied together and the result is stored into D2.
4. Finally the result, i.e. 7448 is displayed on the seven-segment displays.

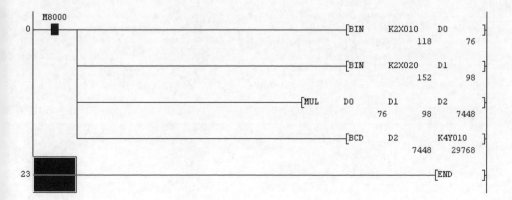

20.4 RPM counter – REV1

The project REV1 counts pulses during a 5-sec timing period and then displays speed in RPM (revs/min).

The pulses are input via X0 and for the purpose of this project are assumed to be obtained from an inductive proximity detector mounted in close proximity to a four-tooth gear wheel connected to a rotating shaft (refer page 200).

The seven-segment display is used to display the speed of the shaft in RPM.

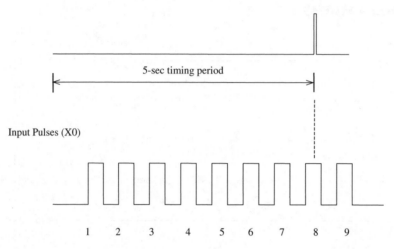

Figure 20.5

From the timing waveforms shown in Figure 20.5, it can be seen that eight pulses are input during the 5-sec timing period.

Ladder diagram – REV1

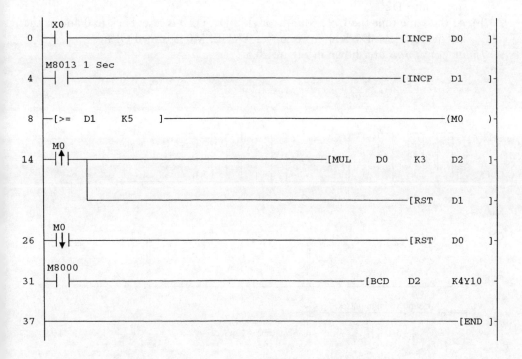

Principle of operation – REV1

1. The simulated input pulses are assumed to be from an inductive proximity detector connected to input X0 and which is mounted in close proximity to a four-tooth gearwheel.
2. Number of pulses obtained in 1 sec $= \dfrac{N}{60} \times 4$ where N = shaft speed in RPM.
   $\qquad\qquad\qquad\qquad\qquad\qquad\qquad\quad$ 4 = no. of gear teeth.

   Number of pulses obtained in 5 sec $= 5 \times \dfrac{N}{60} \times 4 = \dfrac{N}{3}$
   This will be the number of pulses stored in D0 every 5 sec.
3. Therefore $D0 = \dfrac{N}{3}$
4. Hence the shaft speed in RPM (N) $= 3 \times D0$
5. Line 0
   The incoming pulses from the gear wheel, which are input via X0, increment D0.
6. Line 4
   M8013 produces 1-sec clock pulses, which are used to increment D1.
7. Line 8
   (a) The instruction -[>= D1 K5]- will be TRUE, when D1 reaches a count of 5, i.e. after 5 sec.
   (b) Hence every 5 sec M0 is turned ON.

8. Line 14
   (a) On the rising edge of M0, the contents of D0 are multiplied by 3 and the result is stored into D2
   (b) At the same time the 1-sec count stored in D1, i.e. 5 is reset back to 0 and as the instruction -[>= D1 K5]- is no longer TRUE, M0 is turned OFF.
9. *Timing waveforms* are shown in Figure 20.6.

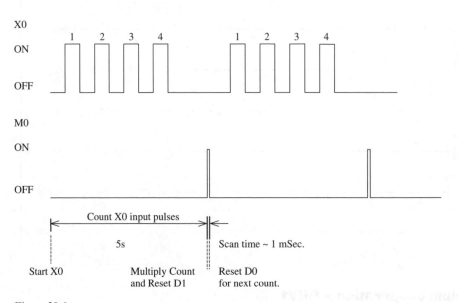

Figure 20.6

10. Line 26
    On the falling edge of M0, i.e. after the 5-sec delay has elapsed, the contents of D0 are reset back to 0.
11. Line 31
    The instruction -[BCD D2 K4Y10]- enables the simulated speed of the shaft to be displayed in RPM on the seven-segment displays.

Monitoring – REV1

1. When D1 is reset back to 0, turn X0 ON and OFF four times to obtain four pulses.
2. Immediately when D1 is reset back to 0, input a further four pulses before D1 reaches 5, i.e. before the next 5 sec have elapsed.

3. The display now becomes as shown below.

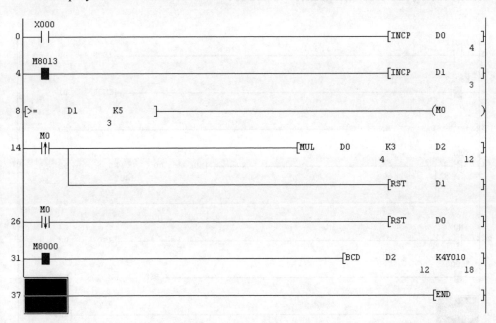

Note

1. The above display was obtained after four pulses were input to X0 during a 5-sec timing period.
2. This value was stored in D0.
3. The contents of D0 were then multiplied by 3 and the result, i.e. 12 was stored in D2.
4. The seven-segment display requires the output of Y10–Y27 to be in BCD format to obtain the value of 12.
   (a) D2 = 12.
   (b) BCD equivalent of 12 in binary (K4Y10) = 0001 0010.
   (c) The decimal equivalent of 0001 0010 = 18.
5. Hence it can be seen, that care must be taken when monitoring BCD Outputs.

REV1A

To prove the accuracy of the system input X0 was replaced with M8011, whose periodic clock time is 0.01 sec, i.e. it produces pulses at a frequency of 100 Hz.

(a) Number of pulses per sec = 100
(b) Number of pulses every 5 sec = 500
(c) Assume the inductive proximity detector also outputs 500 pulses in 5 sec, i.e. D0.
(d) The speed of the rotating shaft in RPM (N) = D0 × 3
$$= 500 \times 3$$
$$= 1500$$

Monitoring ladder diagram – REV1A

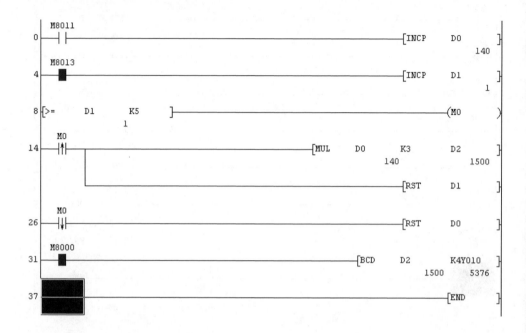

Practical considerations

If a PLC is being considered to actually determine the speed of a rotating shaft, it may be necessary to use the instruction REFF or to use a high speed counter (refer page 247).

20.5 Timing control of a bakery mixer – MIXER1

The project MIXER1 is used in a bakery to control the time taken for mixing dough to the correct consistency.

The project enables the operator to carry out the following:

(a) Set up the mixing time within the range 10–20 sec.
(b) On operating the start button, the output Y0 will be immediately energised for the required mixing time.
(c) The timer values cannot be changed, while the mixer is ON.

Ladder diagram – MIXER1

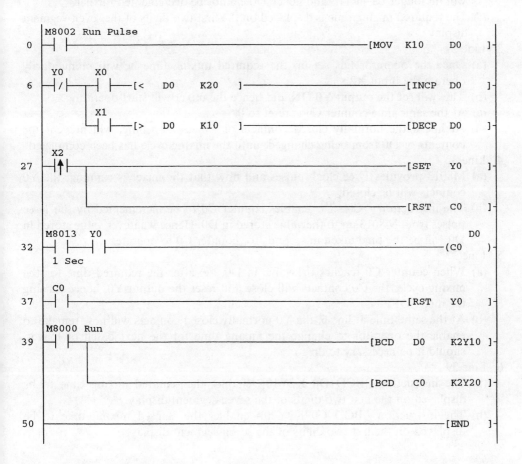

Principle of operation

1. Line 0
   On switching the PLC to RUN, the M8002 contacts will close for one scan time, i.e. approximately 1 milli-sec and this will enable D0 to be loaded with the value 10. The minimum mixing time being 10 sec.
2. Line 6
   (a) The mixer is turned ON and OFF by the output Y0. Hence with Y0 OFF its normally closed contacts will remain closed.
   (b) Opening and closing input X0 will enable D0 to be incremented to a maximum value of 20.
   (c) Once the contents of D0 reach a value of 20, then the instruction -[< D0 K20]- will no longer be TRUE and hence D0 cannot be incremented further.
   (d) Similarly by opening and closing input X1 will enable D0 to be decremented to a minimum value of 10.

(e) Once the contents of D0 reach a value of 10, then the instruction -[> D0 K10]- will no longer be TRUE and hence D0 cannot be decremented further.
(f) The required mixing time is displayed on the first two digits of the seven-segment display.
3. Line 27
    (a) Once the operator has set up the required mixing time he will momentarily operate the input X2.
    (b) This will set the output Y0 ON and hence the mixer will start operating.
    (c) At the same time counter C0 is reset to 0.
    (d) At line 6 the normally closed contacts of Y0 now opening, will prevent the contents of D0 from being changed, until the mixing cycle has been completed.
4. Line 32
    (a) M8013 provides 1-sec clock pulses and now that the mixer is running, the Y0 contacts will be closed.
    (b) The instruction -(C0 D0)- enables counter C0 to be incremented by the 1-sec pulses from M8013 up to the value stored in D0. Hence whatever value stored in D0 will be the time taken in seconds for counter C0 to operate.
5. Line 37
    (a) When counter C0 reaches the value in D0, i.e. after the required time for the mixing cycle, the C0 contacts will close and reset the output Y0, hence turning the mixer OFF.
    (b) At the same time at line 6, the Y0 normally closed contacts will now remake to enable the operator to change the mixing time for the next batch of loaves, should it be necessary to do so.
6. Line 39
    (a) The instruction -[BCD D0 K2Y10]- enables the required mixing time to be displayed on the first two digits of the seven-segment display.
    (b) The instruction -[BCD C0 K2Y20]- enables the elapsed mixing time to be displayed on the last two digits of the seven-segment display.

Monitoring – MIXER1

The monitored display on page 213 shows the following:

(a) The required mixing time as set in D0, is 18 sec.
(b) When the display was obtained, the elapsed mixing time at that point was 10 sec.

Introduction to programs using data registers 213

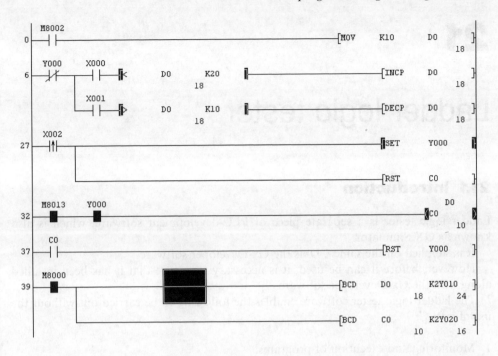

21

Ladder logic tester

21.1 Introduction

Ladder logic tester is a separate piece of PLC-development software, which is also known as GX Simulator.

It is supplied on the same CD as the Gx-Developer software.

However, before it can be used, it is necessary to ensure that it has been installed along with the Gx-Developer software.

The ladder logic tester software enables the following to be carried out without the use of a PLC:

1. Monitoring and execution of programs.
2. Debugging.
3. Timing charts.

The Gx-Developer project is downloaded into the ladder logic tester software, and hence it is the computer, which now effectively executes, monitors and debugs the project instead of the PLC.

21.2 Program execution

To execute a program using ladder logic tester, carry out the following procedure:

1. Open an existing Gx-Developer project, i.e. FLASH1.

```
         X0        T1                                              K10
  0     ─┤ ├──────┤/├──────────────────────────────────────────── (T0    )─

         T0                                                        K10
  5     ─┤ ├──────────────────────────────────────────────────── (T1    )─

         T0
  9     ─┤ ├──────────────────────────────────────────────────── (Y0    )─

 11     ──────────────────────────────────────────────────────── [END   ]─
```

Ladder logic tester 215

2. Select:
   (a) Tools.
   (b) Start ladder logic test.
3. Alternatively, click on the ladder logic test icon (Figure 21.1).

Figure 21.1

4. The display now becomes as shown in Figure 21.2.

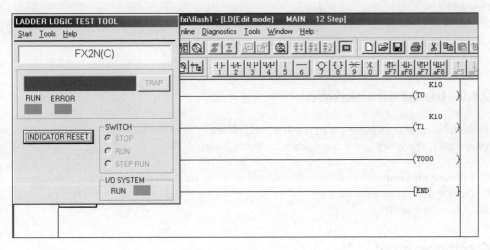

Figure 21.2

5. A few seconds later the displays in Figure 21.3 occur, indicating that the project is being automatically downloaded into the ladder logic test software.

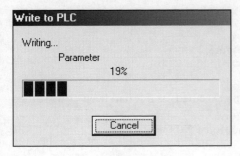 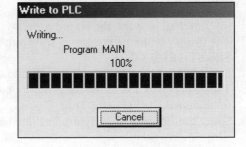

Figure 21.3

6. The display now becomes as shown in Figure 21.4.

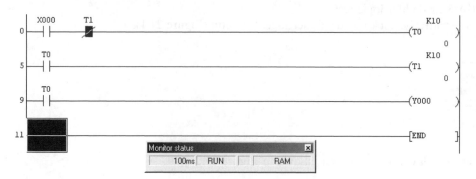

Figure 21.4

7. Note the project is automatically in Monitor Mode.

21.3 Input simulation

To simulate the operation of an input to enable a program to be executed using ladder logic tester, there are three methods which can be used:

1. Forcing an input.
2. Device memory monitor.
3. I/O system settings.

Forcing an input

The Input X0 will now be forced ON using the method described in Section 8.5.

1. Select the following:
   (a) Online.
   (b) Debug.
   (c) Device test.
   (d) Enter x0 into the Bit device window.
2. The display will now appear as shown in Figure 21.5.
3. Select FORCE ON.
4. Select Close to remove the Force Window.
5. The display will now show the program being executed, as if it had been downloaded to an actual PLC (Figure 21.6).
6. Select the Force display again, but this time, force X0 OFF.

Ladder logic tester 217

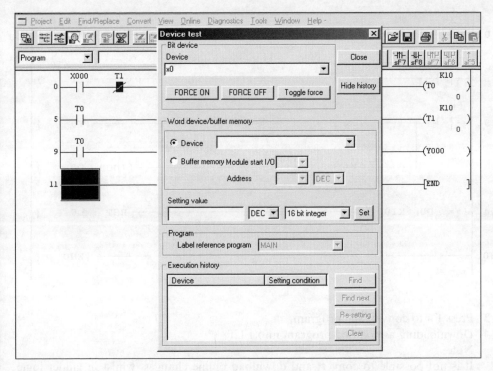

Figure 21.5

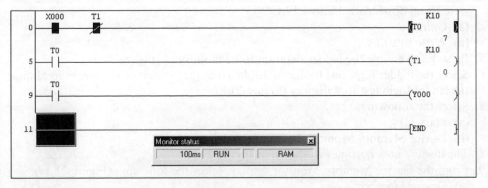

Figure 21.6

21.4 Device memory monitor

Ladder diagram – LLT1

1. Press F2 to leave Monitor Mode.
2. Modify the Ladder Diagram FLASH1 to include Data Register D0 and save it as LLT1.

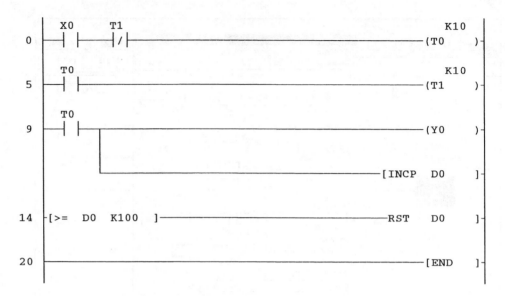

3. Press F4 to convert the program.
4. Downloading a modified program into LLT
   Note:
   It is not possible to convert and download online changes, whilst in ladder logic tester.
   To download LLT1 into ladder logic tester, select the following, as if the program were being downloaded into a PLC:
   (a) Online.
   (b) Write to PLC.
5. Press F3 to enable the ladder diagram to be monitored (Figure 21.7).
6. Select the ladder logic test tool icon displayed at the bottom of the screen to obtain the ladder logic test tool display (Figure 21.8).
7. Select the following:
   (a) Start.
   (b) Device Memory Monitor.
8. The display now becomes as shown in Figure 21.9.
9. From the Device Memory Monitor display, select the following (Figure 21.10):
   (a) Device Memory.
   (b) Bit Device.
   (c) X.
10. The display now becomes as shown in Figure 21.11.
11. Double-click on X0 to force X0 ON. Note the X0 background colour changes to yellow.
12. Move the Device Memory Monitor and the ladder diagram test tool displays downwards, to obtain the overall display shown in Figure 21.12.
13. Select Device Memory again, but this time select Word Device followed by 'D', to monitor the contents of data registers.
14. The display now becomes as shown in Figure 21.13.

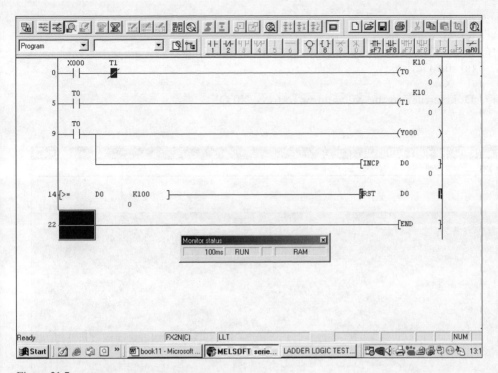

Figure 21.7

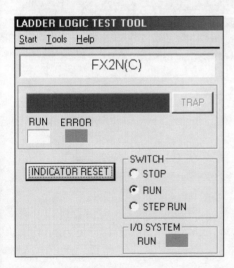

Figure 21.8

15. Note the contents of data registers can be monitored in:
    (a) 16-bit integer.
    (b) 32-bit integer.
    (c) Real.

220 *Mitsubishi FX Programmable Logic Controllers*

   (d) Decimal.
   (e) Hexadecimal.
16. Select Device Memory once and select the following:
   (a) Bit Device.
   (b) X.
17. Double-click in the X0 window to force X0 OFF.

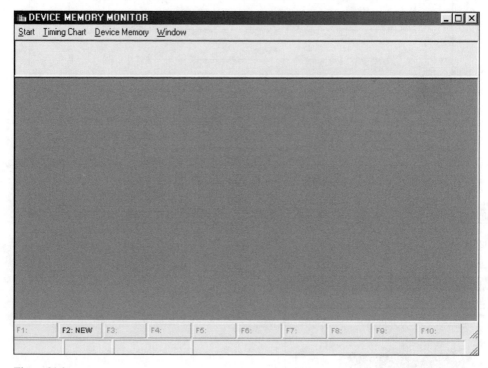

Figure 21.9

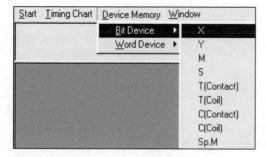

Figure 21.10

Ladder logic tester 221

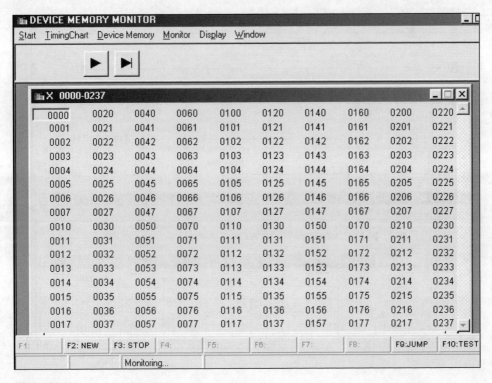

Figure 21.11

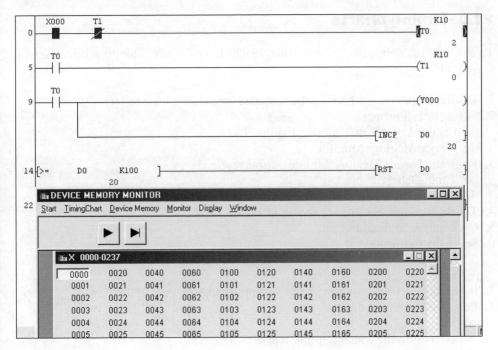

Figure 21.12

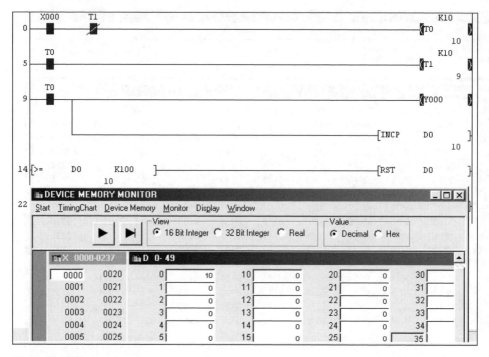

Figure 21.13

21.5 Timing charts

To assist in the debugging of programs, ladder logic tester is able to produce timing waveforms.

1. Ensure the program LLT1 has been loaded into ladder logic tester.
2. Select the following:
   (a) Ladder Logic Test Tool (see Figure 21.8).
   (b) Device Memory Monitor.
3. From the Device Memory Monitor display, select:
   (a) Timing Chart.
   (b) Run.
4. The display now becomes as shown in Figure 21.14.
5. Click on the Manual button.
6. Select the following:
   (a) Device.
   (b) Enter Device.
7. In the Device Entry window, select D, then enter 0 (Figure 21.15).
8. Enter the remaining devices:
   (a) Y0.
   (b) T0 (ensure the contact option is selected).
   (c) X0.

Ladder logic tester 223

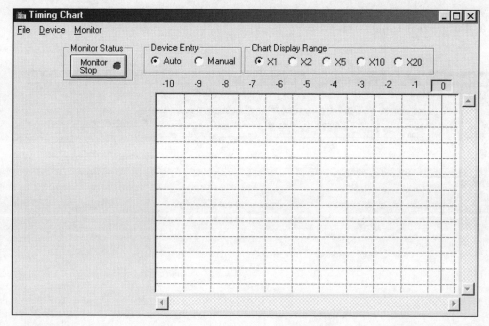

Figure 21.14

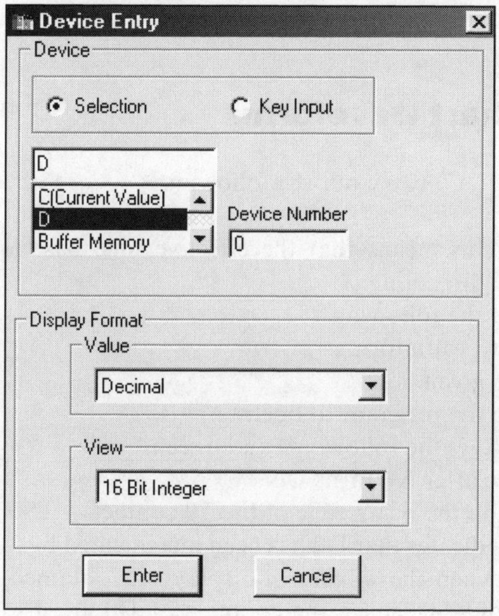

Figure 21.15

9. On selecting Cancel, the display will appear as shown in Figure 21.16.
10. Ensure that Input X0 is OFF.

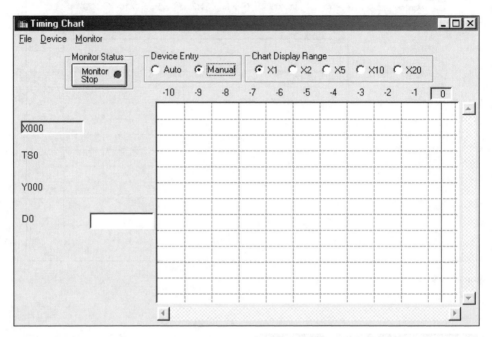

Figure 21.16

21.6 Producing the timing chart waveforms

To obtain the waveforms shown in Figure 21.17, carry out the following:

1. Set the Chart Display Range to X10. This means that the effective time base is 1 sec between each graticule line in the X direction.
2. Select the Monitor Stop button and note the following:
   (a) The wording on the button changes to Monitoring.
   (b) The Monitoring Status light turns to green.
3. Double-click on the X0 button to enable the program to operate.
4. From the waveforms shown in Figure 21.17, the following can be seen:
   (a) The first Y0 output pulse occurs 1 sec after X0 turns ON.
   (b) The contents of D0 are incremented on the rising edge of the T0 contacts – TS0. The cross-over of the D0 waveform indicates that D0 has been incremented by 1.
   (c) From when X0 was turned ON, to when the waveform display was obtained, there have been three rising edges of TS0, hence the contents of D0 are also three.

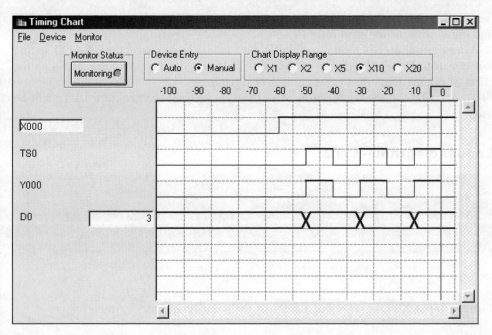

Figure 21.17

21.7 Resetting the timing chart display

To reset and then re-start the display, carry out the following:

1. Double-click on X0.
2. Select D0 and enter 0. This overwrites the existing contents of D0 and resets it back to 0.
3. To restart the program and the display, double-click on X0.

21.8 Saving the setup details

1. To save the setup details, select the following:
   (a) File.
   (b) Save as.
2. Save the Setup file in c:\gxdevel\fxi\LLT1.
3. Select Project.
4. Select Close project. This will close the project LLT1 and also quit ladder logic tester.

21.9 I/O system settings

The I/O system setting option is a very important part of ladder logic tester, in that it enables the simulation of much larger systems. As shown towards the end of this chapter (see Section 21.15), the pneumatic control project PNEU1 can be quite easily simulated and tested without the need for a PLC.

After a program has been downloaded into ladder logic tester, then the I/O setting function will display the effects of forcing inputs and outputs ON/OFF and also the effect of forcing numerical values into data registers.

To describe the basic operation of I/O system setting, the program LLT2 will be used.

Ladder diagram – LLT2

Turning Input X0 OFF/ON will increment the contents of D0.

21.10 Procedure – I/O system setting

1. Quitting ladder logic tester:
   If ladder logic tester is still being executed, then selecting a new project will automatically switch it OFF and return to standard Gx-Developer.
2. Enter the program LLT2.
3. Select ladder logic tester as described in Section 21.2.
4. A few seconds later, the program LLT2 will be downloaded into ladder logic tester.
5. Select Ladder Logic Test Tool from the bottom of the screen (Figure 21.18).
6. The display now becomes as shown in Figure 21.19.
7. Select the following:
   (a) Start.
   (b) I/O System Settings.
8. The display now becomes as shown in Figure 21.20.
9. The I/O Input Settings enables the user to enter conditions, which when are executed will enable the actual ladder diagram input/output requirement, to be forced ON or OFF or for a value to be set into a data register.
10. For example, the first line of the I/O settings will be as shown in Figure 21.21.
11. The display shows that when the condition X0=ON is clicked ON and as the second condition is always ON, then 10 msec later the actual X0 input on the ladder diagram will be Forced ON.
12. Time, in this case, is simulating a 10 msec PLC input delay.

Ladder logic tester 227

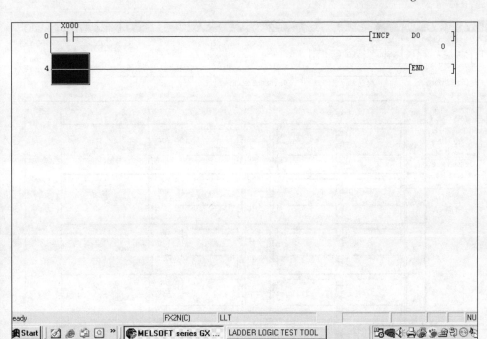

Figure 21.18

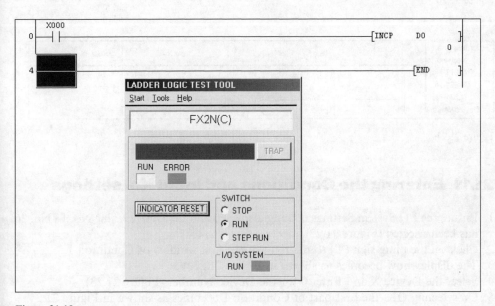

Figure 21.19

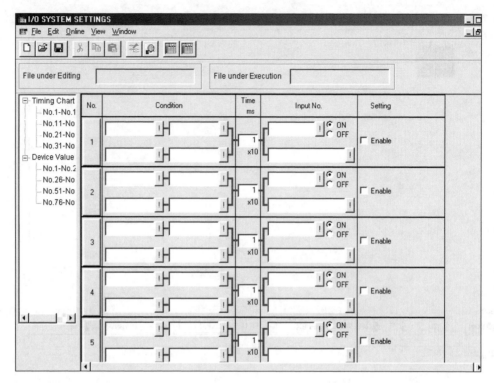

Figure 21.20

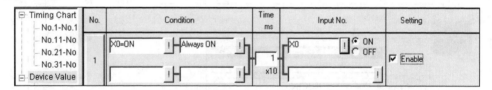

Figure 21.21

21.11 Entering the Conditions and Input No. settings

1. Ensure the I/O System Settings is being displayed and the Device Value No. 1–No. 26 has been selected (Figure 21.22).
2. Click on the pling sign ('!') to the right of the first window of Condition 1.
3. The display now becomes as shown in Figure 21.22.
4. Select the Device X and Enter 0 for the Device Number (Figure 21.23).
5. On selecting OK, the first part of Condition 1 becomes as shown in Figure 21.24.
6. For the second part of Condition 1, select Always ON (Figure 21.25).
7. Hence, Condition 1 now becomes as shown in Figure 21.26.
8. Select the pling (!) for Input 1 to enable the required Bit Device, i.e. Input X0 to be obtained (Figure 21.27).

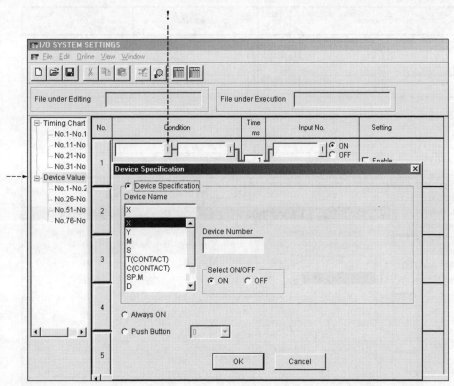

Figure 21.22

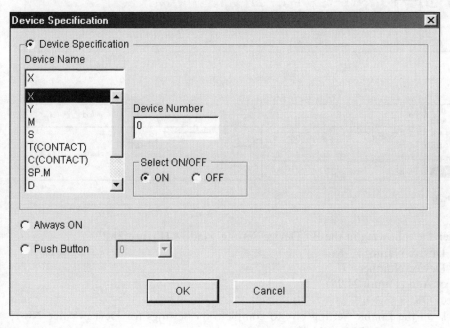

Figure 21.23

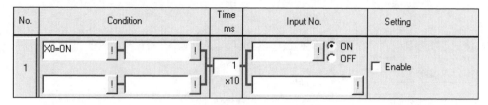

Figure 21.24

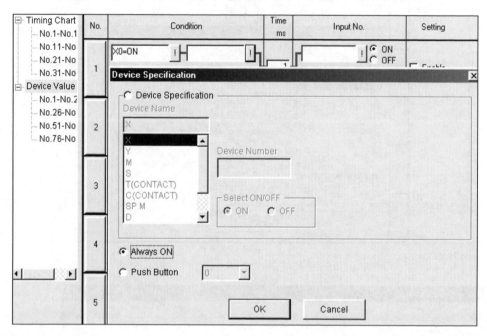

Figure 21.25

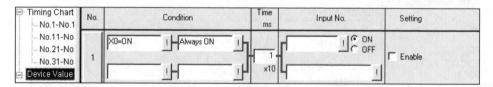

Figure 21.26

9. Enter the following in the Bit Device Setting window (Figure 21.27):
   (a) Device Name   X.
   (b) Device Number   0.
10. Select Add (Figure 21.28).
11. Select OK.
12. Click on the Enable Setting box to complete the settings for Device Value No.1, which is shown in Figure 21.29.

Ladder logic tester 231

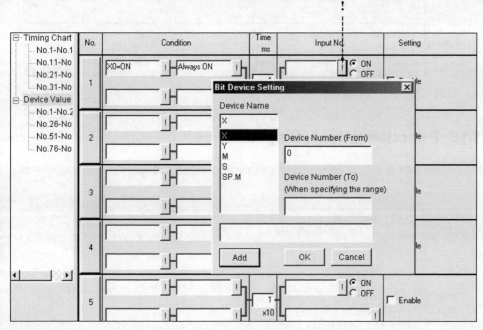

Figure 21.27

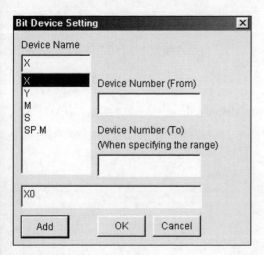

Figure 21.28

13. Save the input/output system file.
14. Select the following:
    (a) File.
    (b) Save As.
15. Save the file as LLT2.IOS in the folder c:\gxdevel\fxi.

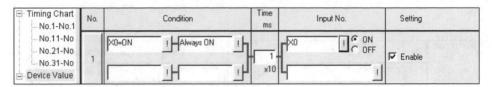

Figure 21.29

21.12 Executing the I/O system

1. From the File menu, select Execute I/O System Settings (Figure 21.30).

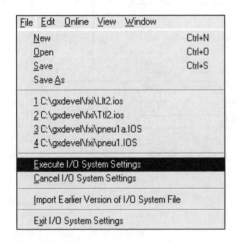

Figure 21.30

2. The window shown in Figure 21.31 is now displayed.

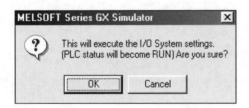

Figure 21.31

3. Select OK.
4. After the I/O system settings have been successfully executed, Select OK once more.
5. To monitor the I/O system, select the following from the I/O System Settings toolbar:
   (a) Online.
   (b) Monitor Mode.
6. The I/O system display now becomes as shown in Figure 21.32.

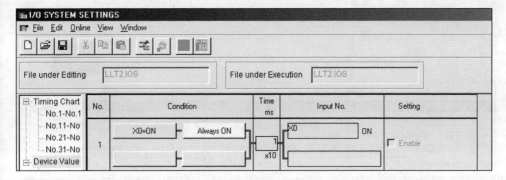

Figure 21.32

7. Move the cursor to the blue border – I/O System Settings – and holding the left-hand mouse button down, drag the display downwards to reveal the Ladder Diagram LLT2 and the the ladder logic test tool displays underneath.
8. Move the I/O system settings and the ladder logic test tool displays downwards, until the overall display becomes as shown in Figure 21.33.

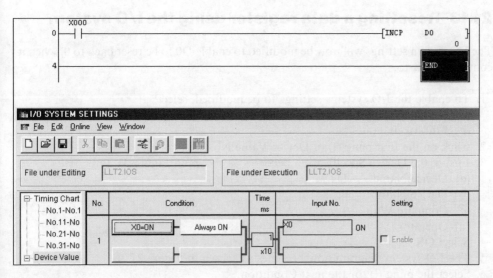

Figure 21.33

9. Click ON the X0=ON window (Figure 21.34).
10. Note the following:
    (a) The X0=ON window background colour has now changed to yellow.
    (b) The X0 contacts have closed.
    (c) The contents of D0 have been incremented to 1.
11. Each time the X0=ON window is selected, the PLC input X0 is toggled ON/OFF.
12. When the X0=ON window is toggled ON, its background colour will change to yellow, and the PLC input X0 will turn ON. This in turn will cause D0 to be incremented.

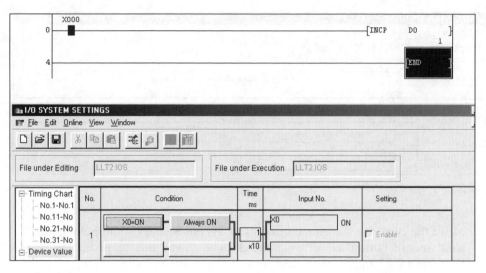

Figure 21.34

21.13 Resetting a data register using the I/O system

The I/O system settings will now be modified to enable D0 to be reset back to 0, when its contents reach 15.

1. To enable the I/O systems settings to be modified, select:
   (a) Online.
   (b) Edit Mode.
2. Click on the first pling (!) of Device Value No. 2.
3. Ensure the Device Specification becomes as shown in Figure 21.35.
   (a) Device Name   D.
   (b) Device Number   0.
   (c) Compare With   15.
   (d) Operator   >=.
4. Select OK.
5. The I/O system diagram now becomes as shown in Figure 21.36.
6. Select the pling (!) for the next Condition.
7. From the Device Specification display, select Always ON.
8. Select the pling (!) for the lower of the two windows of Input 2.
9. The Word Device Setting window now appears (Figure 21.37).
10. Carry out the following:
    (a) Device name   D.
    (b) Device Number   0.
    (c) Setting value   0.
    (d) Select Add.
11. The Word Device Setting now appears as shown in Figure 21.38.
12. Select OK to obtain the modified I/O System Settings diagram (Figure 21.39).

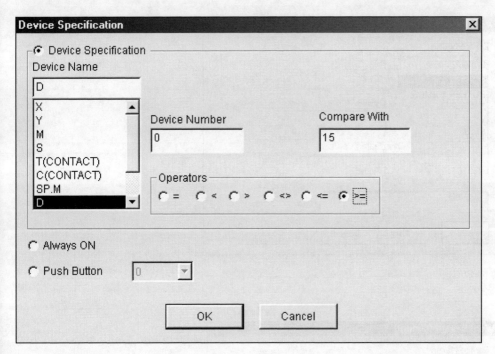

Figure 21.35

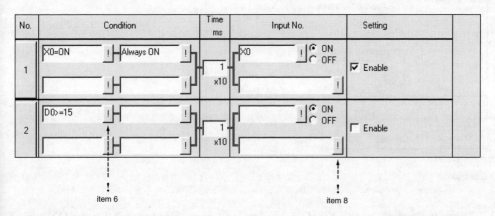

Figure 21.36

13. Finally enable Device Value No. 2 (Figure 21.39).
14. Save the I/O System Setting File LLT2.IOS.
15. From the File menu select Execute I/O System Settings.
16. To monitor the I/O system (Figure 21.39).
    (a) Select Online.
    (b) Monitor Mode.

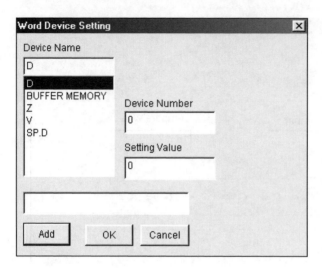

Figure 21.37

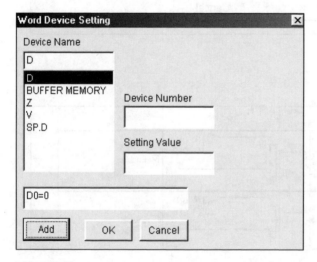

Figure 21.38

17. Repeatedly click on the X0=ON window and each time Input X0 turns ON, then D0 will be incremented.
18. When D0=14, then on the next click of the X0=ON window, D0 will become equal to 15 and the condition D0 >= 15 will become TRUE and hence the I/O system setting will cause D0 to be reset back to 0 (Figure 21.40).

Ladder logic tester 237

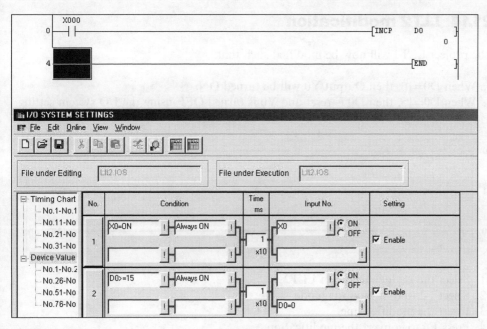

Figure 21.39

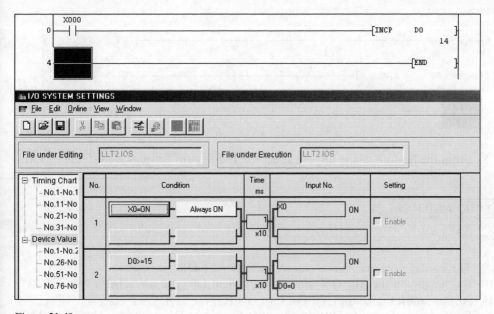

Figure 21.40

21.14 LLT2 modification

The project LLT2 will now be modified such that:

1. When D0=10, then Output Y0 will be turned ON.
2. When D0=15, then D0 is reset and Y0 is turned OFF using the I/O system setting.

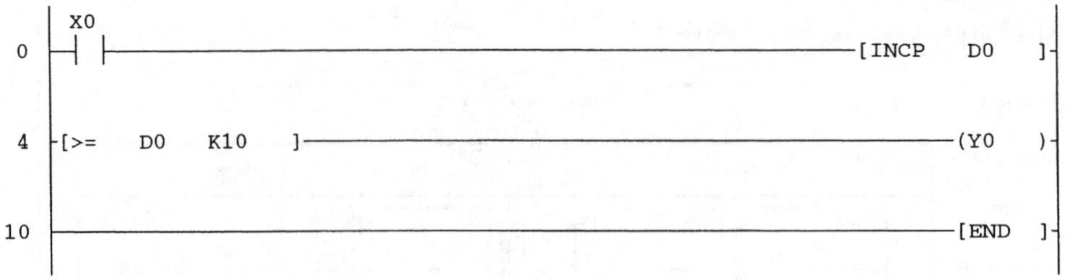

```
      X0
 0   ─┤ ├──────────────────────────────────────────[ INCP    D0   ]─

 4   ─[>=   D0    K10 ]──────────────────────────────────────( Y0  )─

10   ──────────────────────────────────────────────────────[ END   ]─
```

1. Select the Ladder Diagram LLT2.
2. Select F2 to quit monitor mode.
3. Insert the modified line.
4. Press F4 to convert the modification.
5. Download the modified ladder diagram into ladder logic tester.
6. Press F3 to start monitoring.
7. The modified ladder diagram now becomes as shown below:

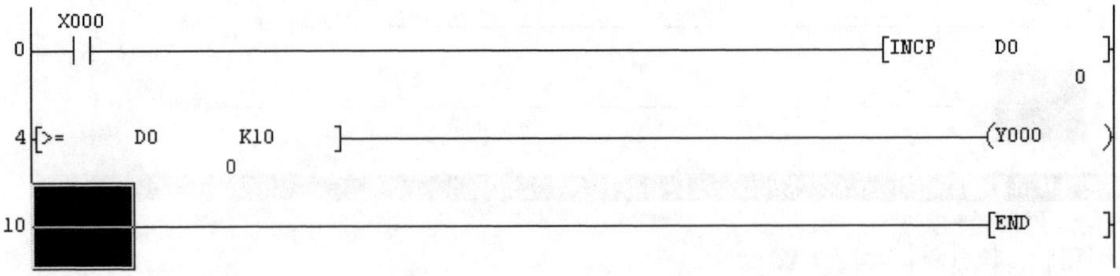

```
      X000
 0   ─┤ ├──────────────────────────────────────────[INCP    D0   ]
                                                              0
 4   [>=   D0    K10 ]──────────────────────────────────(Y000     )
              0
10   ■■■■■─────────────────────────────────────────────[END      ]
```

8. From the bottom of the display select the I/O System Settings icon (Figure 21.41).

```
[▣ I/O SYSTEM SET...]
```

Figure 21.41

9. The display now becomes as shown in Figure 21.42.
10. By clicking on the X0=ON window, increment D0 until it reaches a value of 10.
11. At this point, the Output Y0 will turn ON (Figure 21.43).
12. Continue incrementing D0 and note that as soon as D0 reaches a value of 15 it will immediately be reset and Y0 will turn OFF.

Ladder logic tester 239

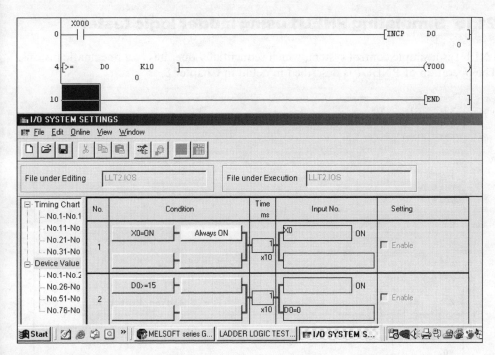

Figure 21.42

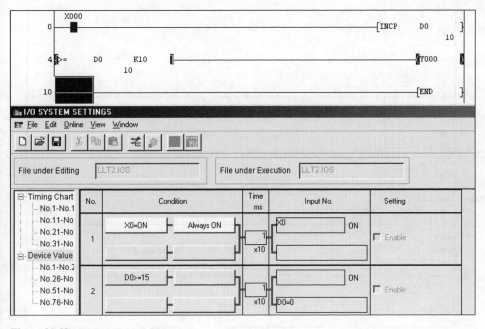

Figure 21.43

21.15 Simulating PNEU1 using ladder logic tester

PNEU1 is a simple control system, which sequentially operates two pneumatic pistons. The operation of PNEU1 is described in detail in Chapter 8.

```
         X0     X1     T0
    0   ─┤├────┤├────┤/├────────────────────────────[MC    N0     M0   ]─
        Start  Stop   5 Sec                                        Start
                      Delay                                        Cycle
         M0
        ─┤├─
        Start
        Cycle


   N0   M0
   ────┤├─────────────────────────────────────────────────────────────────
        Start
        Cycle

         X2     X4     M1                   * < Energise Piston A  >
    7   ─┤├────┤├────┤/├─────────────────────────────────────(Y0   )─
        A-     B-     A&B                                    Sol A
                      FWD
         Y0
        ─┤├─
        Sol A

         X3     X4                          * < Energise Piston B  >
   12   ─┤├────┤├──────────────────────────────────────────(Y1    )─
        A+     B-                                           Sol B
         Y1
        ─┤├─
        Sol B

         X3     X5
   16   ─┤├────┤├──────────────────────────────────────────(M1    )─
        A+     B+                                           A&B
                                                            FWD
         M1
        ─┤├─
        A&B
        FWD

;Timer Section
         M1     X2     X5                                   K50
   20   ─┤├────┤├────┤├──────────────────────────────────(T0     )─
        A&B    A-    B+                                    5 Sec
        FWD                                                Delay

   26   ─────────────────────────────────────────────────[MCR   N0   ]─


   28   ──────────────────────────────────────────────────────[END  ]─
```

21.16 PNEU1 procedure using ladder logic tester

1. Load the project PNEU1.
   This will also cause the end of the ladder logic tester for LLT2.
2. Select Ladder Logic Tester.
3. PNEU1 will automatically be downloaded into ladder logic tester.
4. Select the Ladder Logic Test Tool icon displayed at the bottom of the screen.
5. From the Ladder Logic Test Tool display, select the following:
   (a) Start.
   (b) I/O System Settings.
6. Enter the following settings:
   (a) Device Value No. 1
      (i) X0   Always ON.
      (ii) Y0   M100.   This enables Output Y0 to be monitored. M100 is not used and will always be OFF. This ensures Y0 cannot affect the Input Device X0.
      (iii) Input No.   X0.
   (b) Device Value No. 2
      (i) X1   Always ON.
      (ii) Y1   M100.
      (iii) Input No.   X1.
   (c) Device Value No. 3   X2 Always ON   Input No.   X2.
   (d) Device Value No. 4   X3 Always ON   Input No.   X3.
   (e) Device Value No. 5   X4 Always ON   Input No.   X4.
   (f) Device Value No. 6   X5 Always ON   Input No.   X5.
7. Select all of the Enable boxes for Device Value No's. 1–6.
8. The I/O System Settings now appears as shown in Figure 21.44.
9. Save the I/O System Settings File as PNEU1.IOS.
10. From the File menu now select Execute I/O System Settings.
11. Finally monitor PNEU1.IOS by selecting:
    (a) Online.
    (b) Monitor Mode.

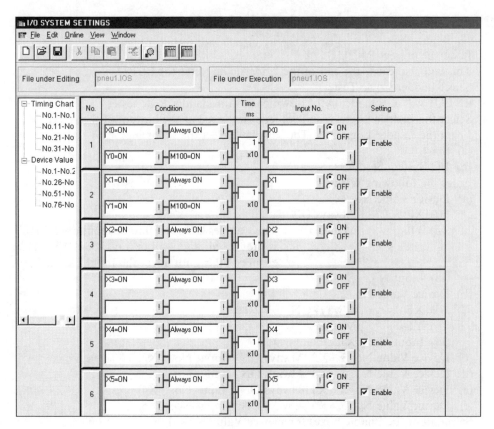

Figure 21.44

21.17 Monitoring procedure

1. With the I/O System Settings now in Monitor Mode, the display will now appear as shown in Figure 21.45.
2. Each time an X0–X5=ON window is selected, the corresponding PLC output will be turned ON/OFF.
3. When an X0 to X5 window is toggled ON, its background colour will change to yellow.
4. Monitor PNEU1.IOS using a procedure similar to the one described on page 98.
   (a) **Display and monitor PNEU1.IOS.**
   (b) **Turn ON Conditions X1, X2 and X4.**
       This will simulate the operation of the Stop push button and the A– and B– limit switches.
   (c) **Turn ON and then Turn OFF Condition X0.**
       This simulates the momentary operation of the Start push button.
       This will cause Y0 to turn ON. In a PLC-controlled system Piston A would now move to the A+ position and therefore limit switch X2 would open and X3 would close.

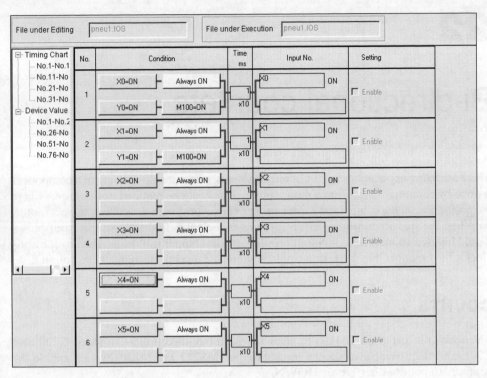

Figure 21.45

(d) **Turn OFF X2 and Turn ON X3.**
This will cause Y1 to turn ON. In a PLC controlled system, Piston B would now move to the B+ position and therefore limit switch X4 would open and X5 would close.

(e) **Turn OFF X4 and Turn ON X5.**
Auxiliary Output M1 will energise and cause Output Y0 to turn OFF. Hence Piston A will return to its A− position.

(f) **Turn OFF X3 and Turn ON X2.**
With X2 closing, this will start the operation of the Timer T0. After 5 seconds, Timer T0 will time out and its normally closed contact (Line 0) will open, breaking the master control circuit of M0. Output Y1 will now turn OFF.

(g) **Turn X5 OFF and Turn ON X4.**

(h) **Turn X0 ON then OFF,** to repeat the process.

22

Bi-directional counters

There are many applications in PLC control systems, where the positioning of a component is done by counting pulses from a datum point. This could be required, for example, where the positioning of the component is done using a conveyor system driven by a stepper motor.

When the stepper motor is driven forward in a clockwise direction, the input pulses would be used to increment a counter and when the stepper motor is pulsed in the anti-clockwise direction, the same input pulses would be used to decrement the counter.

COUNT5

The following program COUNT5 uses the 32-bit bi-directional counters, C200 and C201, and their associated special memory coils, M8200 and M8201, which enable the counters to count either UP or DOWN.

Counter C200 is set to a positive value, i.e. K10 and Counter C201 is set to a negative value, i.e. K−10.

22.1 Ladder diagram – COUNT5

```
       X0   START
  0    ─┤ ├──────────────────────────────[DMOVP   K-11    C200 ]─
                                         
                                         ─────────────────[DMOVP   K-11    C201 ]─

       X1   UP/DOWN
 19    ─┤ ├──────────────────────────────────────────────────(M8200 )─
                                         
                                         ────────────────────────(M8201 )─

       X2   COUNT                                                K10
 24    ─┤ ├──────────────────────────────────────────────────(C200  )─
                                                                 K-10
                                         ────────────────────────(C201  )─

       C200  C201
 35    ─┤/├──┤ ├─────────────────────────────────────────────(Y0    )─


 38    ──────────────────────────────────────────────────────[END ]─
```

22.2 Special memory coils M8200–M8234

The special memory coils M8200–M8234 are only used in conjunction with the bi-directional counters C200–C234, respectively.

Each of these special memory coils control whether or not their associated bi-directional counters will count UP or count DOWN.

Special memory coil	Associated bi-directional counter
M8200	C200
M8201	C201
"	"
"	"
"	"
"	"
"	"
M8234	C234

22.3 Principle of operation – COUNT5

1. Line 0
   (a) The operation of X0 will load both 32-bit counters with the value −11.
   (b) Since the counters are 32-bit, it is necessary to use the DMOV Instruction.
2. Line 19
   (a) Input X1 operates the Special Memory Coils M8200 and M8201.
   (b) When M8200 and M8201 are not operated, then the Counters C200 and C201 will count UP.
   (c) Conversely, when M8200 and M8201 are operated, then the Counters C200 and C201 will count DOWN.
3. Line 24
   (a) The pulsing of Input X2 will simultaneously increment or decrement both Counters C200 and C201.
   (b) Input X1 is used to determine the direction of the count.
4. Line 35
   The Output Y0 will be ON, while the count is in the range −10 to +9.

22.4 Operating procedure

1. Momentarily, operate Input X0 to enable both counters to be loaded with the value −11.
2. Ensure Input X1 is OFF, so that both counters can be incremented.
3. Pulse Input X2 to increment both counters to +10.
4. Turn Input X1 ON, so that both counters can be decremented.
5. Pulse Input X2 to decrement both counters to −11.

22.5 Monitoring – COUNT5

Monitor counters C200 and C201 and the Output Y0 to confirm the following (Figure 22.1):

1. Incrementing count
   When incrementing the count:
   (a) C200 will turn ON, when the count goes from +9 to +10.
   (b) C201 will turn ON, when the count goes from −11 to −10.
2. Decrementing count
   When decrementing the count:
   (a) C200 will turn OFF, when the count goes from +10 to +9.
   (b) C201 will turn OFF, when the count goes from −10 to −11.
3. Output Y0
   The Output Y0 will be ON, while the count is in the range −10 to +9.

Device	ON/OFF/Current	Setting value	Connect	Coil	Device comment
X000			0		
X001			1		
X002			0		
C200(D)	−10	10	0	0	
C201(D)	−10	--------	1	0	
Y000			1		

Figure 22.1

23
High-speed counters

23.1 Introduction

A high-speed counter would be used, when it is necessary to count a high-frequency input signal, which is coming, for example, from an encoder disk.

These high-frequency signals can be counted by feeding the signals into the Inputs X0–X5 and using the high-speed 32-bit Counters C235–C255.

Using either Input X0 or X1 it is possible to use to count input signals whose frequency can be up to 60 kHz, while the other inputs X2–X5 can count input signals up to 10 kHz.

All X Inputs, other than X0–X7, have a fixed input filter time of 10 msec, whilst each of the Inputs X0–X7 has a programmable input filter, which can be adjusted from 50 μsec to 60 msec, using the Instruction REFF.

However, when using standard counters, i.e. C0–C234 there also is the scan time of the program to be considered, which depending on the size of the program, can be from a few milli-seconds to tens of milli-seconds.

Therefore, although the input filter time can be reduced, the scan time cannot and hence if is required to count a signal whose input frequency is greater than a few tens of hertz, an alternative method is required, i.e. high-speed counters.

High-speed counters use a method known as interrupts.

When the logic level of a high-frequency input signal goes positive, the present instruction completes its execution requirements and then the program scan is immediately halted, i.e. interrupted, and the contents of the corresponding high-speed counter are incremented by one.

After this has been done, the program returns to its normal scan.

23.2 Types of high-speed counters

Within the FX range of PLCs there are four types of high-speed counters, all of which are 32-bit and which can be used as up/down counters.

The available types are:

1. Single phase with user start/reset (C235–C240)
   This is the simplest type of high-speed counter in that it has a single input and can be reset from the main program by using any X input, except the one designated as the high-speed input.

2. Single phase with assigned start/reset (C241–C245)
   This type of counter is reset by using assigned X inputs. The advantage of this type of counter is that the reset input is interrupt-driven and hence does not have to wait for the normal scan of the program, before the counter can be reset. In addition, two of these counters have separate start inputs.
3. Two-phase bi-directional (C246–C250)
   This type of counter requires at least three assigned X inputs:
   (a) Count up.
   (b) Count Down.
   (c) Reset – except C246.
   In addition, two of these counters also have separate start inputs. For example, consider C247.
   (a) When pulses are applied to X0, then C247 will count UP.
   (b) When pulses are applied to X1, then C247 will count DOWN.
   (c) When +24 V is applied to X2, then C247 is Reset.
4. A/B phase (C251–C255)
   This type of counter is used with an encoder disk that has two tracks, A and B, which are misaligned by 90 degrees to one another. The purpose of the A/B phase system is to enable a high-speed counter to count UP, when the disk is rotating in one direction, and count DOWN, when rotating in the other direction. As shown in the encoder drawing, when rotating anti-clockwise, the leading edge of the A signal occurs while the B phase signal is LOW. This will cause the counter to count UP. Similarly, when the encoder disk is rotating clockwise, the leading edge of the A signal occurs, while the B signal is HIGH. This will cause the counter to count down.

Encoder disk

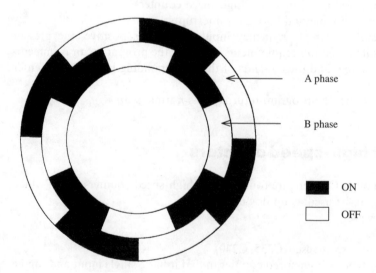

Figure 23.1

Encoder waveforms

Up-count

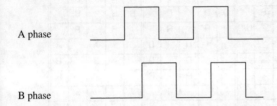

A phase

B phase

Down-count

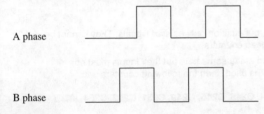

A phase

B phase

Figure 23.2

Encoder accuracy

The drawing of the encoder disk shows that it only has four ON/OFF segments per cycle. Hence, for every revolution, there are only four complete pulses. This gives an accuracy of $360/4 = 90°$.

Industrial encoders

A modern industrial encoder can produce 16384 pulses per revolution, which means that the position of the shaft driving the encoder disk, can be measured to an accuracy of $360°/16384 = 0.022°$.

23.3 FX range of high-speed counters

The high-speed counters C235–C255, listed in Figure 23.3, can be used with the following PLCs: FX1S, FX1N, FX2N, FX2NC.

INPUT	1 Phase counter user start/reset						1 Phase counter assigned start/reset					2 Phase counter bi-directional					A/B Phase counter				
	C235	C236	C237	C238	C239	C240	C241	C242	C243	C244	C245	C246	C247	C248	C249	C250	C251	C252	C253	C254	C255
X0	U/D						U/D			U/D		U	U		U		A	A		A	
X1		U/D					R			R		D	D		D		B	B		B	
X2			U/D					U/D			U/D	R		R			R		R	
X3				U/D				R		R			U		U			A		A
X4					U/D				U/D				D		D			B		B
X5						U/D			R				R		R			R		R
X6											S				S					S	
X7											S					S					S

Key:
U – up counter input D – down counter input
R – reset counter (input) S – start counter (input)
A – A phase counter input B – B phase counter input

 – Counter is backed up/latched

Input assignment:

- X6 and X7 are also high speed inputs, but function only as start signals. They cannot be used as the counted inputs for high-speed counters.
- Different types of counters can be used at the same time but their inputs must not coincide. For example, if counter C247 is used, then the following counters and instructions cannot be used;
  C235, C236, C237, C241, C242, C244, C245, C246, C249, C251, C252, C254, I0□□, I1□□, I2□□.

Counter Speeds:
- General counting frequencies:
  - Single-phase and bi-directional counters; up to 10 kHz.
  - A/B phase counters; up to 5 kHz.
  - Maximum total counting frequency (A/B phase counter count twice) FX1S & FX1N 60 kHz, FX2N & FX2NC 20 kHZ.

- For FX2N & FX2NC Inputs X0 and X1 are equipped with special hardware that allows higher-speed counting as follows:
  - Single-phase or bi-directional counting (depending on unit) with C235, C236 or C246; up to 60 kHz.
  - Two phase counting with C251; up to 30 kHz.

Figure 23.3

23.4 High-speed counter inputs

Unlike standard counters, i.e. C0–C234, the high-speed counters C235–C255 are assigned specific inputs, i.e. X0–X5.

It is important to note that when using high-speed counters, the corresponding X input does *not* appear on the ladder diagram.

Single-phase counter inputs

Input	Counter
X0	C235
X1	C236
X2	C237
X3	C238
X4	C239
X5	C240

Hence, if a high-frequency signal were connected to X0, then the signal would increment/decrement Counter C235.

Similarly, if the signal were connected to X5, this would increment/decrement Counter C240.

23.5 Up/down counting

To enable the single-phase high-speed counters to count either UP or DOWN is done by operating an associated memory coil.

Input	High-speed counter	Up/down memory coil
X0	C235	M8235
X1	C236	M8236
X2	C237	M8237
X3	C238	M8238
X4	C239	M8239
X5	C240	M8240

That is,

1. When M8235 is OFF, then Counter C235, counts UP.
2. When M8235 is ON, then Counter C235, counts DOWN.

23.6 Selecting the high-speed counter

Selecting the high-speed counter and its associated X input is done by turning the Counter Coil, i.e. C235 ON using the M8000 run contact, as shown below:

```
       │ X0
    0  ├──┤ ├──────────────────────────────────────[RST   C235  ]─
       │
       │ X7
    3  ├──┤ ├──────────────────────────────────────────(M8235  )─
       │
       │ M8000 RUN                                              K10
    6  ├──┤ ├──────────────────────────────────────────(C235   )─
       │
       │ C235
   12  ├──┤ ├──────────────────────────────────────────(Y0     )─
       │
   14  ├────────────────────────────────────────────────[END   ]─
```

As Counter C235 has been selected, i.e. its coil has been reserved for high-speed counting, then each time a high-speed pulse is applied to Input X0, Counter C235 will be incremented.

It can be seen from the above diagram that Input X0 is *not* shown, as it is not scanned during the execution of the ladder diagram program.

When ten pulses have been applied to Input X0, Counter C235 will operate and its contacts can then be used to drive an output, i.e. turn Y0 ON, or execute an instruction.

23.7 Maximum total counting frequency

If all of the high-speed inputs, X0–X5, have been assigned to be used with high-speed counters, then when any of these inputs goes positive, the main program scan will be interrupted while the corresponding high-speed counter is incremented/decremented.

Program control is then passed back to the main program.

Therefore, if all of the above inputs were operating at their maximum frequencies, it would be impossible to scan and execute the main program.

Hence, for the FX2N and the FX2NC PLCs a limit of 20 kHz is placed on the sum of the individual frequencies of every signal being applied to the high-speed inputs.

For the FX1S and the FX1N, the limit is 60 kHz.

Example

Using either an FX2N or FX2NC PLC

1. Let the frequency of the input signal to X0 be 10 kHz.
2. Let the frequency of the input signal to X1 be 7 kHz.
3. The total counting frequency is therefore 17 kHz.

Since the total counting frequency is less than the maximum of 20 kHz, the above example could be used.

Note

The inputs X0 and X1 of the FX2N and FX2NC PLCs have a special hardware which enables higher counting frequencies than the other inputs X2–X5.

1. Single-phase maximum counting frequency – 60 kHz.
2. A/B phase maximum counting frequency – 30 kHz.

All of the counting frequencies are reduced, if the high-speed Instructions HSCS, HSCR and HSZ are used.
Please refer to the Mitsubishi FX Programming Manual II for further details.

23.8 High-speed counter – HSC1

Block diagram

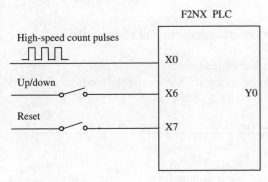

Figure 23.4

Ladder diagram – HSC1

```
      X6   UP/DOWN
 0   ─┤├──────────────────────────────────────────────(M8235  )─
      X7   RESET
 3   ─┤├──────────────────────────────────────[RST   C235   ]─
      M8002
     ─┤├─
      M8000                                                K10
 7   ─┤├──────────────────────────────────────────────(C235   )─
                                              SELECT C235
      C235
13   ─┤├──────────────────────────────────────────────(Y0     )─
      M8000
15   ─┤├──────────────────────────────────[DBCD  C235  K4Y10 ]─
                                              DISPLAY COUNT
25                                                    ─[END  ]─
```

Note

1. High-speed counter reset
   It is essential that all high-speed counters are initially reset, before being used.
2. Switch bounce
   If a switch is used to input pulses into X0, then connect a 0.1 µF capacitor from the X0 input terminal to 0 V. This ensures switch bounce pulses are not counted.

Principle of operation

1. Line 0
   (a) Operating Input X6 will energise the up/down Memory Coil M8235.
   (b) If M8235 is OFF, then the counter will count UP. If M8235 is ON, then the counter will count DOWN.
2. Line 3
   (a) The operation of X7 will reset the high-speed Counter C235.
   (b) On switching to RUN, M8002 will operate for one scan time, and this will also reset the counter.
3. Line 7
   (a) On switching to RUN, M8000 will operate and this will energise the coil of the high-speed Counter C235.
   (b) This then selects Input X0 to be a high-speed input.
   (c) With M8235 OFF, pulses are now applied to X0 to enable Counter C235 to count UP.
4. Line 13
   When the input count reaches 10, then the contacts of C235 will operate to energise Output Y0.
5. Line 0 – counting down
   Operate X6 to energise M8235 and note the following:
   (a) Operating Input X0 will now cause the count in C235 to decrement.
   (b) When the count goes from 10 to 9, then the C235 contacts (Line 13) will open and turn the Output Y0 OFF.
6. Line 15
   (a) The Instruction -[DBCD C235 K4Y10]- will enable the contents of the 32-bit high-speed Counter C235 to be monitored on the 7-segment displays.
   (b) Note:
   If the counter value increments by more than one, each time Input X0 is operated, then this shows that the counter is counting the high-speed switch bounces, each time the switch contacts close (see Figure 23.4).

23.9 Decade divider – HSC2

The purpose of the program HSC2 is to demonstrate the use of the two high-speed compare Instructions DHSCR and DHSZ. These instructions enable the high-speed input pulses to be divided by ten and also for the resulting lower frequency output to have an equal Mark/Space ratio.

When using high-speed counters, it is not possible to use standard comparison instructions to determine when such counters reach a particular value. This is because the counters operate using interrupts, whereas the standard comparison instructions are executed during the normal scan of the program.

Ladder diagram – HSC2

```
       X7
  0    ─┤├──────────────────────────────────────────[RST    C235  ]─
       M8002
       ─┤├──┘
       M8000                                                   K30
  4    ─┤├───────────────────────────────────────────────────(C235  )─
        │
        └────────────────────────────[DHSCR   K10    C235    C235  ]─
       M8000
 23    ─┤├──────────────────[DHSZ    K0     K4     C235     M0    ]─
       M2
 41    ─┤├───────────────────────────────────────────────────(Y0   )─
       M8000
 43    ─┤├──────────────────────────────[DBCD    C235    K4Y10    ]─

 53    ─────────────────────────────────────────────────────[END  ]─
```

Note

1. The FX1S, FX1N and the FX2N PLCs can only use a maximum of six DHSCS/R and DHSZ Instructions within a program.
2. If a switch is used to input pulses into X0 then, as in HSC1, connect an 0.1 µF capacitor from the X0 input terminal to 0 V. This ensures switch bounce pulses are not counted.

High-speed counter reset – DHSCR

This instruction is used to reset a specified output, when the value in the high-speed counter also reaches a specified value.

The DHSCR Instruction used in the program is -[DHSCR K10 C235 C235]-.

When the count in C235 reaches the value of 10, then C235 is reset back to 0. This means that the contents of C235 will always be in the range 0–9, i.e. a decade counter.

High-speed zone compare – DHSZ

This instruction is used to compare the contents of a high-speed counter and set three consecutive outputs, depending on the counter's contents.

The DHSZ Instruction used in the program is -[DHSZ K0 K4 C235 M0]-.

If the contents of C235 are:

1. Less than 0   M0 is turned ON.
2. Greater than 0 but less than or equal to 4   M1 is turned ON.
3. Greater than 4   M2 is turned ON.

When the count reaches a value of 5, M2 and hence Y0 turn ON.

When the count reaches a value of 10, high-speed Counter C235 is reset back to 0 and therefore M2 and Y0 turn OFF.

Therefore:

1. Between a count of 0 and 4, Y0 is OFF.
2. Between a count of 5 and 10, Y0 is ON.

Waveforms

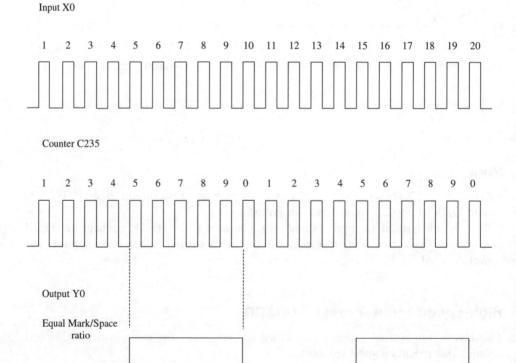

Figure 23.5

Principle of operation – HSC2

1. Line 0
   Input X7 or Memory Coil M8002 – RUN pulse is used to reset C235.
2. Line 4
   (a) The coil of the high-speed Counter C235 is energised and hence Input X0 is automatically assigned as the high-speed input. The valve of K30 is a dummy value, which will never be reached, as the counter is always reset at a value of 10.
   (b) Input pulses can now be applied to Input X0 and C235 will count up.
   (c) When the count in C235 reaches a value of 10, then C235 itself resets back to 0.
3. Line 23
   (a) When the count in C235 reaches 5, then M2 will turn ON, since the count is greater than 4.
   (b) M2 will stay ON until the counter reaches 10 and then the counter is reset back to 0.
   (c) With the count at 0, then since this is less than 4, M2 will turn OFF.
4. Line 41
   As M2 turns ON and OFF, so will the Output Y0.

Monitor

Monitor the following using entry data monitor (Figure 23.6).

1. X0
2. C235
3. M0
4. M1
5. M2
6. Y0

Device	ON/OFF/Current	Setting value	Connect	Coil	Device comment
X000			0		
C235(D)	7	30	0	1	
M0			0		
M1			0		
M2			1		
Y000			1		

Figure 23.6

23.10 Motor controller – HSC3

This project simulates the operation of either an AC or DC motor having a single-phase encoder disk connected to its output shaft.

The pulses from the encoder disk are counted using a high-speed counter.

The number of pulses in the high-speed counter are then used to control the speed and the position of the motor.

Motor speed waveform

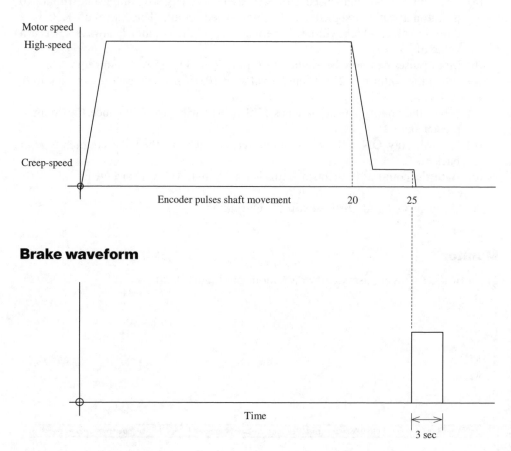

Figure 23.7

Sequence of operation

1. The encoder pulses will be input to the PLC via X0 and will be counted using the high-speed Counter C235.
2. On operating the Start Input X1, the motor accelerates up to its high-speed setting.
3. After counting twenty encoder pulses, the motor decelerates to its creep-speed setting.
4. After counting twenty-five encoder pulses, the creep-speed is turned OFF and a brake is applied to the motor for 3 sec. This ensures the motor shaft is stopped at the required position.

Ladder diagram – HSC3

```
                                          *<X0 - High Speed Input    >
        M8000                                                    K999
  0     ─┤├─────────────────────────────────────────────────────(C235  )─
        X7
  6     ─┤├──┬────────────────────────────────────────────[RST   C235  ]─
        T0  │
        ─┤├─┘

        X1   Y1    Y2
 10     ─┤├──┤/├───┤/├────────────────────────────────────────────(M0   )─
       Start Creep Brake
        M0
        ─┤├──┘

        M0                                *<High Speed ON           >
 15     ─┤├─────────────────────────────────────────────────[SET  Y0   ]─

        M8000                             *<High Speed OFF          >
 17     ─┤├──┬───────────────────────────[DHSCR   K20     C235    Y0   ]─
             │                            *<Creep Speed ON          >
             ├───────────────────────────[DHSCS   K20     C235    Y1   ]─
             │                            *<Creep Speed OFF         >
             ├───────────────────────────[DHSCR   K25     C235    Y1   ]─
             │                            *<Brake ON                >
             ├───────────────────────────[DHSCS   K25     C235    Y2   ]─
             │                            *<Display Count           >
             └────────────────────────────────[DBCD       C235    Y10  ]─

                                          *<Brake Timer             >
        Y2                                                       K30
 79     ─┤├─────────────────────────────────────────────────────(T0    )─
       Brake

        T0                                *<Brake OFF               >
 83     ─┤├────────────────────────────────────────────────[RST    Y2  ]─

 85     ─────────────────────────────────────────────────────────[END  ]─
```

Principle of operation

1. Line 0
   C235 is selected as the high-speed counter.
2. Line 6
   Counter C235 is reset by either the manual operation of Input X7, or when Timer T0 times out at the end of each complete cycle of the motor's speed and position.
3. Line 10
   The operation of Input X1 will operate M0, provided the motor is not in creep speed or the brake is ON. M0 will now latch over its own contact.
4. Line 15
   The closure of M0 will set the Output Y0 and this will cause the motor to start and accelerate up to its high-speed setting. The motor will now rotate at high speed.
5. While the motor is running, the encoder will be outputting pulses to the Input X0, where they are being counted by the high-speed Counter C235.
6. Line 17
   (a) When the number of pulses received from the encoder reaches twenty, this indicates that the motor shaft has almost reached its required position and the motor will now decelerate to its creep speed.
   (b) The Instruction -[DHSCR K20 C235 Y0]- will reset Output Y0, and hence switch OFF the high-speed drive to the motor.
   (c) The Instruction -[DHSCS K20 C235 Y1]- will set Output Y1, and hence switch ON the creep-speed drive to the motor.
   (d) When the number of pulses received from the encoder reaches twenty-five, this indicates that the motor shaft has now reached its required position.
   (e) The Instruction -[DHSCR K25 C235 Y1]- will reset Output Y1, and hence switch OFF the creep-speed drive to the motor.
   (f) The Instruction -[DHSCS K25 C235 Y2]- will set Output Y2, and this will activate the brake solenoid and hence bring the motor to an immediate stop.
7. Line 79
   The closure of the Y2 contacts starts the 3-sec timer T0.
8. Line 83
   After 3 sec, the timer will time out and reset Y0. This will cause the brake solenoid to become de-energised.
9. Line 6
   The closure of the T0 contacts will also reset the high-speed Counter C235.

23.11 A/B phase counter – HSC4

The object of this project is to simulate an A/B phase up-down, high-speed Counter (see Figure 23.2).

To simulate the A/B encoder signals, two waveforms were produced and output via the Y0 and Y1 outputs.

These output waveforms have the same clock frequency, i.e. 0.5 Hz, but have a phase difference of 90° to one another.

High-speed counters 261

The output pulses were then connected into the X0 and X1 inputs, which will be used as the A/B inputs for the high-speed Counter C251.

Tests were carried out with both the relay and the transistor versions of the FX2N PLC.

It was found that because of the low-frequency clock pulses, which were used as input signals into X0 and X1, any high-frequency switch bouncing of the FX2N relay outputs had no effect on the A/B high-speed Counter.

Wiring diagram

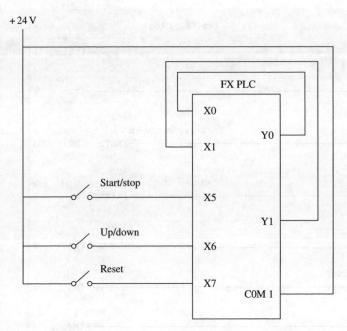

Figure 23.8

Ladder diagram – HSC4

```
                                          *<A/B Phase Counter      >
       M8000   RUN                                          K999
  0    ─┤ ├──────────────────────────────────────────────(C251   )─
       X7
  6    ─┤ ├──────────────────────────────────[RST     C251    ]─
       M8002
       ─┤ ├─

       START                                *<1 Sec Clock          >
       X5      T1                                             K5
 10    ─┤ ├───┤/├─────────────────────────────────────────(T0    )─

       T0                                                    K5
 15    ─┤ ├───────────────────────────────────────────────(T1    )─

       T0                                   *<2 Sec A Clock        >
 19    ─┤/├──────────────────────────────────────[ALTP    M0    ]─

       T0                                   *<2 Sec B Clock        >
 23    ─┤ ├──────────────────────────────────────[ALTP    M1    ]─

       M0      X6   COUNT UP
 27    ─┤ ├───┤/├─────────────────────────────────────────(M2    )─
               X6   COUNT DOWN
              ─┤ ├────────────────────────────────────────(M3    )─

       M1      X6   COUNT UP
 34    ─┤ ├───┤/├─────────────────────────────────────────(M4    )─
               X6   COUNT DOWN
              ─┤ ├────────────────────────────────────────(M5    )─

       M2                                   *<A/B Phase Output 1   >
 41    ─┤ ├───────────────────────────────────────────────(Y0    )─
       M5
       ─┤ ├─

       M3                                   *<A/B Phase Output 2   >
 44    ─┤ ├───────────────────────────────────────────────(Y1    )─
       M4
       ─┤ ├─

 47    ──────────────────────────────────────────────────[END    ]─
```

Principle of operation

1. Line 0
   (a) The closure of M8000 on switching to run will select the A/B phase high-speed Counter C251.
   (b) Since C251 has been selected, then Input X0 will be used as the A phase input, and Input X1 will be used as the B phase input.
2. Line 10
   (a) The Input X5 is used to start/stop the count pulses.
   (b) Timers T0 and T1 are used to produce 1 sec equal Mark/Space ratio signals.
3. Lines 19 and 23
   (a) On starting up, the T0 normally closed contacts will be closed and hence the Instruction -[ALTP M0]- will turn M0 ON.
   (b) On the next closure of the T0 normally closed contacts, M0 will be turned OFF.
   (c) Now the T0 normally open contacts will not close for a further 0.5 sec.
   (d) When the T0 normally open contacts do close, the Instruction -[ALTP M1]- will now turn M1 ON 0.5 sec after M0, i.e. a phase lag of 90°.
   (e) On the next closure of the T0 normally open contacts, M1 will be turned OFF.
   (f) Hence, M0 and M1 produce two waveforms, whose periodic times are both 2 sec. However, M0 leads M1 by 90°.

A/B phase waveforms

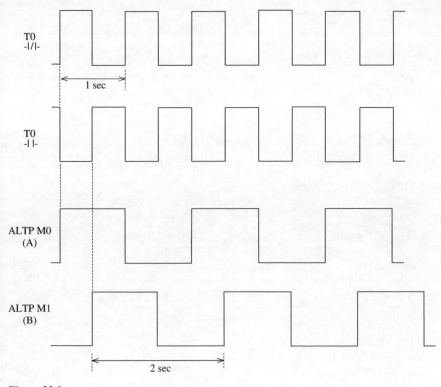

Figure 23.9

4. Lines 27 and 34
   (a) Input X6 is used to enable the A/B high-speed Counter C251 to either count up or count down.
   (b) With X6 not operated, then the 2-sec A clock will pulse M2 – count up.
   (c) Similarly, the 2-sec B clock will pulse M4 – count Up.
   (d) If X6 is operated, then the A clock will pulse M3 – count down
   (e) Similarly the B clock will pulse M5 – count down.
5. Lines 41 and 44
   (a) M2 and M4 will pulse Outputs Y0 and Y1, and these Outputs have been connected to the Inputs X0 and X1, respectively.
   (b) Since Y0 leads Y1 by 90°, then the high-speed Counter C251 will count up.
   (c) When X6 is operated, then M5 and M3 will now pulse Outputs Y0 and Y1, respectively.
   (d) Since Y0 now lags Y1 by 90°, then C251 will count down.

24

Floating point numbers

Floating point number operation is available only with the FX2N and FX2NC PLCs, and not with the FX1S and FX1N PLCs.

If these PLCs are not available, then the floating point programs in this section can still be evaluated by using ladder logic tester.

24.1 Floating point number range

The advantage of using floating point operation in PLCs is that very small and very large numbers can be processed.

Number range $-3.403 \times 10^{+38}$ to $+3.403 \times 10^{+38}$
Smallest available change $+/-1.175 \times 10^{-38}$

24.2 Number representation

Consider the result of 22/7. This can be represented using three different numbering methods:

1. Integer   Result = 3; remainder = 1; i.e. no decimal values.
2. Floating point   Result = 3.1428.
3. Scientific notation Result = 31428×10^{-4}.

24.3 Floating point instructions

The following instructions are used with integer, floating point and scientific notation numbers.

FLT

1. Converts an integer number to a floating point number, when M8023 is OFF.
2. Converts a floating point number back to an integer number, when M8023 is ON.

This instruction, like INT, converts the floating point number to the rounded down equivalent integer number.

INT

Converts a floating point number to a 16-bit integer number, irrespective of the state of M8023.

DINT

Converts a floating point number to a 32-bit integer number, irrespective of the state of M8023.

DEBCD

Converts a floating point number to a scientific notation number.

DEBIN

Converts a scientific notation number to a floating point number.

DEADD

Adds two floating point numbers together. However, if one or both of the numbers are an integer constant they are automatically converted to floating point.

DEMUL

Multiplies two floating point numbers together. However, if one or both of the numbers are integer constants they are automatically converted to floating point.

24.4 Storing floating point numbers – FLT1

The following program, stores the floating point number 6.5 in Data Registers D2 and D3.

```
         M8000
   0      ┤ ├─────────────────────────────[MOV    K65    D0  ]

                                          [MOV    K-1    D1  ]

                                          [DEBIN  D0     D2  ]

   20                                                   [END ]
```

Principle of operation

1. Line 0
   (a) The Instructions -[MOV K65 D0]- and -[MOV K-1 D1]- will store the number 6.5 in scientific format in D0 and D1, i.e. $6.5 = 65 \times 10^{-1}$

Floating point numbers 267

(b) The Instruction -[DEBIN D0 D2]- will now convert the scientific format number into floating point format and store the result into two Data Registers D2 and D3.
2. Note:
All floating point numbers are stored in two consecutive data registers, i.e. they are stored in 32-bit binary, using a format recommended by the American Institute of Electronic and Electrical Engineers – IEEE.

24.5 Monitor – ladder diagram FLT1

Download the program FLT1 into an FX2N PLC and then press F3 to monitor.

As can be seen, the 32-bit number stored in D2 and D3, i.e. 1087373312 at this moment bears no relationship to 6.5.

24.6 Device batch monitoring

1. Monitor the contents of D0, D1, D2 and D3 using device batch.
2. Monitor the registers in 16-bit integer format (Figure 24.1).

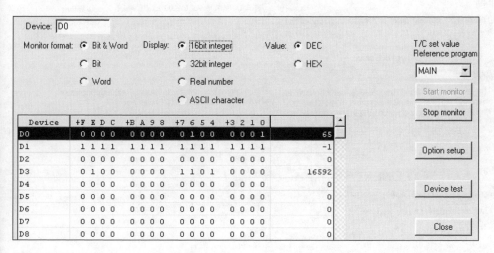

Figure 24.1

3. The contents of the data registers are:
   (a) D0   0000 0000 0100 0001 = 65.
   (b) D1   1111 1111 1111 1111 = −1.
   (c) D2   0000 0000 0000 0000 = 0.
   (d) D3   0100 0000 1101 0000 = 16592.
4. As mentioned earlier, the numbers stored in D2 and D3 still bear no relationship as yet to the number 6.5. This is because floating point numbers are stored in a special format recommended by the IEEE.

24.7 Floating point format

Introduction

To allow calculations with very small and very large numbers the FX PC uses floating point numbers. The format used in the FX PC has been taken from the recommendations of I.E.E.E (Institute of Electrical and Electronic Engineers) for the application of floating point in personal and micro computers.

Floating point

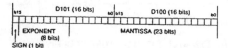

The diagram above shows the bit map for a floating point number. The number is stored in two consecutive data registers and has a sign, mantissa and exponent.

In principal the actual number is calculated by

$$\pm \text{Mantissa} \times 2^{\text{Exponent}}$$

The multiplying factor is 2 because the values are held in binary.

Sign

The sign for floating point is only an indicator of the sign.

   0 = positive value
   1 = negative value

Exponent

The exponent is an 8 bit positive number that is interpreted to give an actual exponent of −126 to +127. This is achieved by subtracting the value 127 from the

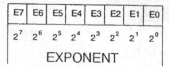

actual value of the exponent.

Thus the exponent has a value of:

$$\left(E7 \times 2^7 + E6 \times 2^6 + \ldots + E0 \times 2^0\right) - 127$$

Figure 24.2

Mantissa

The mantissa is a 23 bit positive binary number. The values represent 23 binary places and the number is assumed to be preceded by a 1.

Thus the mantissa has a value of:

$$1 \times 2^0 + A22 \times 2^{-1} + A21 \times 2^{-2} + \ldots + A0 \times 2^{-23}$$

Example

D101 = 16592   D100 = 0
     = 40D0 HEX   = 0000 HEX

```
  4    0    D    0    0    0    0    0
0110000010 1101 0000000000000000000000
 | EXPONENT |          MANTISSA
SIGN
```

The sign is set to 0; positive.

The exponent is set to 10000001 which is equivalent to

$$\left(1 \times 2^7 + 0 \times 2^6 + \ldots + 1 \times 2^0\right) - 127$$

$$= \left(128 + 0 + \ldots + 1\right) - 127$$

$$= 2$$

The mantissa is set to 10100000000000000000000 which is equivalent to 1.101 binary or

$$1 \times 2^0 + 1 \times 2^{-1} + 0 \times 2^{-2} + 1 \times 2^{-3} + \ldots + 0 \times 2^{-23}$$

$$= 1 + \frac{1}{2} + \frac{0}{4} + \frac{1}{8} + \ldots + \frac{0}{8388608}$$

$$= 1.625$$

Therefore the number equals

$$+ 1.625 \times 2^2 = 6.5$$

24.8 Obtaining the floating point value

From the monitored results in Section 24.6 of the binary patterns for D2 and D3, the floating point value can be obtained.

D3																D2															
b31	b30			b27	b26				b23	b22			b19			b16	b15							b8	b7						b0
0	1	0	0	0	0	0	0	0	1	1	0	1	0	0	0	0	0	0	0	0	0	0	0	0	0	0	0	0	0	0	0
P	Exponent									Mantissa																					

The floating point value is stored in 2 × 16-bit Data Registers D2 and D3 as follows:

1. The Mantissa is stored in bits b0–b22.
2. The Exponent is stored in bits b23–b30.
3. The Polarity is stored in bit 31; 0 positive number; 1 negative number.

The weighting value of each binary bit is shown in the table below except for bits b15 to b0, which in this instance are all at 0.

Parity	Exponent								Mantissa							
b31	b30			b27	b26			b23	b22			b19				b16
	128	64	32	16	8	4	2	1	1/2	1/4	1/8	1/16	1/32		1/64	1/128
0	1	0	0	0	0	0	0	1	1	0	1	0	0		0	0

The floating point value is determined from the following equation:

Floating point value = Mantissa × 2^{Exponent}

The exponent (binary) = 1000 0001
= 129
= 129 − 127 = 2 The 127 is deducted automatically.

Exponent value = 2^2 = 4 The Power calculation is done automatically.

Mantissa value = 1 + 1/2 + 1/8
= 1.625 The addition of the initial 1 is also done automatically.

Floating point value
= 4 × 1.625
= 6.5

24.9 Device batch monitoring – floating point numbers

1. As described in Section 24.6, use device batch to monitor the contents of Data Registers D0, D1, D2 and D3.
2. Select the option Real number, to obtain the display in Figure 24.3.

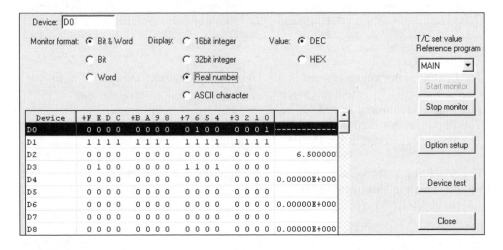

Figure 24.3

3. As can be seen from the display, the Real number (Floating Point) equivalent of the contents of D2 and D3 is 6.5.

24.10 Area of a circle – FLT2

This program now uses floating point instructions to determine the area of a circle using the formula Area $= \pi r^2$, where r is an incrementing/decrementing value in mm.

It also shows how an accurate integer result can be obtained to the nearest whole number. If $r = 2\,mm$, then the area of the circle is $12.568\,mm^2$.

The Instruction INT will now round down this floating point result to $12\,mm^2$.

However, if 0.5 is added to the float result, this produces an answer of $13.068\,mm^2$, and the Instruction INT will now round down this value to the nearest correct Integer value of $13\,mm^2$.

24.11 Ladder diagram – FLT2

```
Flt2
Area of a circle
        M8002
   0 ───┤ ├──────────────────────────────────────[RST   D0  ]
        X0    X1   M8013
   4 ───┤ ├──┤/├──┤↑├─────────────────────────────[INC   D0  ]
                  (1 Sec)
        X1    X0   M8013
  11 ───┤ ├──┤/├──┤↑├────[>   D0   K0 ]──────────[DEC   D0  ]

Convert radius and PI to floating point
        M8000                         *<PI to Scientific Format
  23 ───┤ ├───────────────────────────────[MOV    K0     D1  ]
          │
          │                                 ─────[MOV    K3142  D2  ]
          │
          │                              *<3142 exp-3 = 3.142
          │                                 ─────[MOV    K-3    D3  ]
          │
          │                              *<Float radius
          │                                 ─────[DEBIN  D0     D4  ]
          │
          │                              *<Float D2, D6 = 3.142
          │                                 ─────[DEBIN  D2     D6  ]

* Calculate area of circle
        M8000                            *<Radius squared
  57 ───┤ ├───────────────────────[DEMUL  D4     D4     D8  ]
          │
          │                              *<Result1 = 3.142 x(r squared)
          │                          ───[DEMUL  D6     D8     D10 ]

* Convert Result to nearest integer
        M8000
  84 ───┤ ├───────────────────────────────[MOV    K5     D12 ]
          │
          │                                 ─────[MOV    K-1    D13 ]
          │
          │                              *<5 exp-1 = 0.5
          │                                 ─────[DEBIN  D12    D14 ]
          │
          │                              *<Add 0.5 to result1
          │                          ───[DEADD  D10    D14    D16 ]
          │
          │                              *<Integer result
          │                                 ─────[INT    D16    D18 ]


 122 ────────────────────────────────────────────────────[END ]
```

Floating point numbers 271

24.12 Principle of operation – FLT2

1. The object of the program FLT2 is to determine the area of a circle whose radius is an incrementing/decrementing value in mm. Then, depending on the float result, convert it to the nearest whole integer value.
2. Lines 4
   (a) Operating just Input X0 enables the radius 'r', which is stored in D0, to be incremented by 1 mm every second.
   (b) If Inputs X0 and X1 are both operated, then D0 can neither increment nor decrement.
3. Line 11
   (a) Operating just Input X1 enables D0 to be decremented every second.
   (b) The Instruction -[> D0 K0]- ensures that the radius cannot be decremented below zero.
4. Line 23
   (a) Moving the value K0 into D1 produces a scientific value in D0 and D1 which is equal to $r \times 10^0$ which is equal to r.
   (b) The scientific value of π stored in D2 and D3 is 3142×10^{-3}.
   (c) The Instruction -[DEBIN D0 D4]- converts the scientific value of 'r' stored in D0 and D1 into floating point format, which is then stored in D4 and D5.
   (d) The Instruction -[DEBIN D2 D6]- converts the scientific value of π, i.e. 3142×10^{-3} into floating point format, which is then stored in D6 and D7.
5. Line 57
   (a) The instruction -[DEMUL D4 D4 D8]- enables the floating point value of the radius to be multiplied by itself, i.e. r^2, and the result is stored in D8 and D9.
   (b) The Instruction -[DEMUL D6 D8 D10]- enables the floating point value of the area of the circle, i.e. πr^2 to be obtained and the result is stored in D10 and D11.
6. Line 84
   (a) This block of instructions is used to add 0.5 to float result1 to determine the more accurate integer result.
   (b) For example, let r = 2 mm.

Symbol	Value	Data registers
r integer	2	D0
r float	2.00	D4 D5
r^2	4.00	D8 D9
πr^2	12.568	D10 D11
Constant	0.5	D14 D15
$0.5 + \pi r^2$	13.068	D16 D17
INT($0.5 + \pi r^2$)	13	D18

   (c) The Instruction -[INT D16 D18]- will produce in D18 an integer value of 13, which is the nearest integer value to 12.568.

24.13 Monitored results – FLT2

1. Download FLT2 to the FX2N PLC.
2. Press F3 to monitor the ladder diagram.
3. Operate Input X0 until D0 becomes 2.
4. Confirm the following results:
   (a) D0 = 2   r
   (b) D8 = 4.00 r^2
   (c) D10 = 12.568 πr^2 Area of circle to three decimal places.
   (d) D16 = 13.068 $0.5 + \pi r^2$
   (e) D18 = 13   Area of circle to the nearest whole integer.
5. Operate Input X0 until D0 becomes 4.
6. Confirm the following results:
   (a) D0 = 4   r
   (b) D8 = 16 r^2
   (c) D10 = 50.272 πr^2 Area of circle to three decimal places.
   (d) D16 = 50.772 $0.5 + \pi r^2$
   (e) D18 = 50   Area of circle to the nearest whole integer.

24.14 Floating point – ladder logic tester

Using ladder logic tester (see Chapter 21), it is possible to test the program FLT2, without the use of an FX2N PLC.

1. Use entry ladder monitoring (refer Chapter 12) so that the bottom of Line 84 and Line 4 can be displayed together.
2. Select Ladder Logic Tester.
3. Select IOS System Settings and enter the simulated Inputs X0 and X1.
4. Save the IOS System Settings file as flt2.ios.
5. Arrange the display until it is as shown in Figure 24.4.
6. Operate the X0=ON button until D0=4 and note that when the radius of a circle is equal to 4 mm, its area to the nearest integer is 50 mm^2.

274 Mitsubishi FX Programmable Logic Controllers

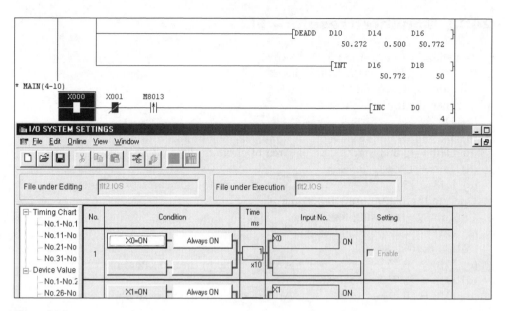

Figure 24.4

25

Master control – nesting

This section on the master control instruction is in addition to the program MC1, which is described on page 87.

The master control instruction enables a group of instructions to be controlled by a single master contact, which effectively switches power to those instructions.

The MCR instruction is used at the end of the set of instructions and any instructions following the MCR instruction are not under the control of the master contact.

25.1 Nesting level

The nesting level number refers to the number of a master control circuit, which is linked within other master control circuits.

As can be seen from Figure 25.1, the level 1 instructions are contained or nested within the level 0 instructions.

By using nested master control instructions, sets of instructions can be switched in or out of the operating sequence, depending on the requirements of the process.

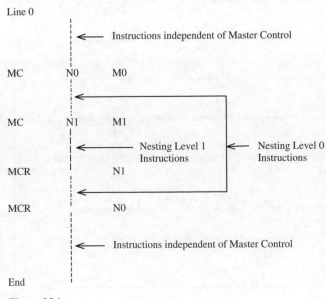

Figure 25.1

25.2 Ladder diagram – MC2

```
         M8013
    0    ─┤├─────────────────────────────────────────[INCP  D0    ]─

         M8000
    4    ─┤├─────────────────────────────────────────[BCD   D0   K2Y10]─
*Start Master Control
          X0    X1
   10    ─┤├───┤├──────────────────────────────────[MC    N0    M0  ]─
          │
          M0    │
         ─┤├────┘
*Nesting level 0
N0      M0
──┬──────┤├───
  │
  │      T1                                                      K10
  │  16  ─┤/├──────────────────────────────────────────────────(T0   )─
  │
  │      T0                                                      K10
  │  20  ─┤├───────────────────────────────────────────────────(T1   )─
  │
  │      T0
  │  24  ─┤├───────────────────────────────────────────────────(Y0   )─
  │
  │      X2
  │  26  ─┤├───────────────────────────────────[MC    N1    M1  ]─
*Nesting Level 1
N1      M1
──┬──────┤├───
  │
  │      T3                                                      K20
  │  30  ─┤/├──────────────────────────────────────────────────(T2   )─
  │
  │      T1                                                      K20
  │  34  ─┤├───────────────────────────────────────────────────(T3   )─
  │
  │      T2
  │  38  ─┤├───────────────────────────────────────────────────(Y1   )─
*End Nesting Level 1
   40   ──────────────────────────────────────────────[MCR   N1   ]─

         M8013
   42   ─┤├──────────────────────────────────────────────────(Y6   )─
*End Nesting Level 0
   44   ──────────────────────────────────────────────[MCR   N0   ]─
*Independent of Master Control
         M8013
   46   ─┤├──────────────────────────────────────────────────(Y7   )─

   48   ──────────────────────────────────────────────────────[END ]─
```

Note: The master control contacts shown on the ladder diagram MC2 *cannot* be entered from the keyboard. They appear on the diagram, after it has been converted to an instruction program.

25.3 Principle of operation

1. Lines 0 and 4
   (a) Lines 0 and 4 occur before the master control instruction and hence they will operate as normal.
   (b) Each time the 1-sec clock M8013 pulses ON then data register D0 will be incremented by 1 and the contents of D0 will be displayed on the seven-segment displays.
2. Line 10
   (a) Operation of the start push button X0 will operate the instruction -[MC N0 M0]-, which will cause the following to occur.
   (b) Memory coil M0 will be energised and latch over its own contact.
   (c) The operation of M0 will now energise the master control contact M0.
   (d) This effectively enables all of the nesting level 0 instructions from line 16 to line 26 to become operative.
3. Lines 16–24
   (a) The two timers T0 and T1 are interconnected to form a 1-sec ON, 1-sec OFF circuit.
   (b) The contacts of T0 will now continuously turn the output Y0 ON for 1 sec and then OFF for 1 sec.
4. Line 26
   (a) The operation of input X2, will now energise the nesting level 1 instructions, which are controlled by master control contact M1.
   (b) This now energises a 2-sec ON/OFF circuit.
5. Lines 30–38
   (a) The two timers T2 and T3 are interconnected to form a 2-sec ON, 2-sec OFF circuit.
   (b) The contacts of T2 will now continuously turn the output Y1 ON for 2 sec and then OFF for 2 sec.
6. Note: For the nesting level 1 instructions to become operative, both master control contacts M0 and M1 have to be ON.
7. Lines 42 and 44
   (a) Line 42 continues the nesting level 0 instructions and hence the remaining Level 0 instructions will be executed until the reset instruction -[MCR N0]- is executed.
   (b) The instructions at line 42 use the 1-sec M8013 clock to turn the output Y6 ON/OFF.
8. Lines 44 and 46
   (a) The instruction -[MCR N0]- terminates the master control sections and the program returns to standard ladder diagram operation.
   (b) The instructions at line 46 use the 1-sec M8013 clock to turn the output Y6 ON/OFF independently of the master control instructions.

Monitoring – MC2

Using the ladder diagram, monitor the operation of MC2.

(a) With no input switches ON.
(b) Operating master control level 0.
(c) Operating master control levels 0 and 1.

26
Shift registers

A Shift register is a combination of memory elements, which are linked together to store data and then shift the stored data in a leftwards or rightwards direction.

26.1 Shift register applications

The following is a list of applications, which require the use of a shift register.

(a) Parallel to serial conversion.
(b) Serial to parallel conversion.
(c) Storage and shifting of data relating to industrial processes.
(d) Scanning of input signals sequentially, i.e. multiplexing.

26.2 Basic shift register operation

The basic principle of operation for the ladder diagram shown below, is that memory coils M27–M20 are combined to become 1 × 8 bit shift register.

The input data is determined by the logic state of input X0.

The shift right signal is obtained from the input X1, i.e. whenever X1 is operated, the logic state of X0 is shifted into the msb (M27) of the shift register and the existing contents of the shift register are shifted one place right (see Figure 26.1).

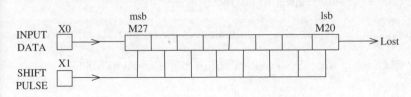

Figure 26.1

26.3 Ladder diagram – SHIFT1

```
       X1
0  ────┤ ├──────────────────────────[SFTRP   X0    M20    K8    K1 ]─
       SHIFT                                 INPUT
       INPUT                                 DATA

       M8000
10 ────┤ ├──────────────────────────────[MOV      K2M20  K2Y0     ]─
       │                                                 LED DISPLAY
       │
       └───────────────────────────────[MOV       K2M20  D0       ]─
                                                         SHIFT REG.
                                                         MONITOR

       X2
21 ────┤ ├──────────────────────────────[ZRST     M20    M27      ]─
       RESET

27 ─────────────────────────────────────────────────────────[END ]─
```

26.4 Principle of operation – SHIFT1

1. Line 0
   Each time input X1 is operated, the following occurs:
   (a) The contents of the shift register M27–M20 are shifted one position right.
   (b) The logic state of X0, i.e. a logic 0 or 1 is loaded into the msb of the shift register, i.e. M27.
2. Line 10
   (a) The contents of the shift register are displayed on the LED Outputs Y0–Y7.
   (b) Transferring the contents of the shift register to D0 will enable the individual bit contents of the shift register to be monitored using device batch.
3. Line 21
   (a) The operation of input X2 will enable the instruction -[ZRST M20 M27]- to be executed.
   (b) The zone reset instruction resets all of the M coils from M20 to M27, which therefore causes the shift register to be reset.

Note:
   (a) When a logic 1 in the lsb, i.e. M20 is shifted out of the register, its value becomes lost.
   (b) However when using rotate instructions, the output is saved in M8022.

26.5 Operating procedure

1. Monitor the contents of the shift register using device batch (refer Section 26.6).
2. Set in turn, each of the X0 conditions as shown below, in the serial input data column.
3. Then operate the shift button X1 just once and note the corresponding change in the contents of the shift register.

Shift register contents

Serial input data (X0)	M27 (Y7)	M26 (Y6)	M25 (Y5)	M24 (Y4)	M23 (Y3)	M22 (Y2)	M21 (Y1)	M20 (Y0)
lsb 1	1	0	0	0	0	0	0	0
0	0	1	0	0	0	0	0	0
0	0	0	1	0	0	0	0	0
1	1	0	0	1	0	0	0	0
0	0	1	0	0	1	0	0	0
1	1	0	1	0	0	1	0	0
1	1	1	0	1	0	0	1	0
msb 0	0	1	1	0	1	0	0	1 = 105 decimal

                 msb   lsb

The serial input data 0 1 1 0 1 0 0 1 is converted by the shift register to parallel data, with an equivalent decimal value of 105.

26.6 Monitoring – SHIFT1

Using device batch, monitor the contents of D0 for each of the X0 conditions, as shown in the previous table.

Figure 26.2 shows the contents of the shift register after the last shift pulse.

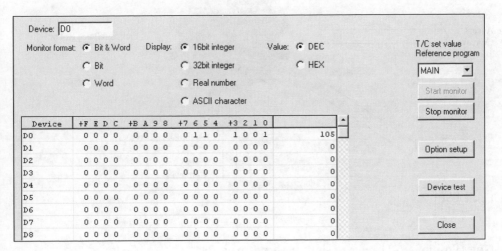

Figure 26.2

27

Rotary indexing table

The use of rotary indexing tables is found in many industrial processes and readily lends itself to being controlled by shift registers.

The following application could be used in a food processing plant, which produces jars of pickled onions, and requires that the lid be screwed tightly on to ensure the pickled onions are hermetically sealed (see Figure 27.1).

This section describes how a lid can be automatically placed on top of the jar and then screwed on.

27.1 Index table system – plan view

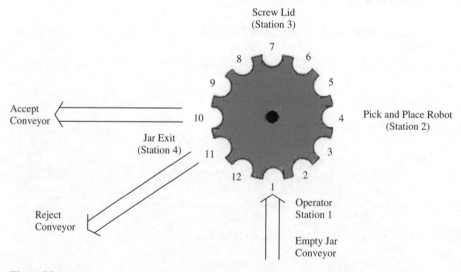

Figure 27.1

Note: Positions 2–11 refer to shift register 1, page 284.

27.2 System requirements

To meet the system requirements, the following design was formulated.

1. At station 1, a sensor connected to X0 is used to detect that an empty jar had been inserted into the rotary index table.
2. The jar is then filled with pickled onions.
3. The operator on observing that a jar is at station 1 and that it has been filled with pickles will operate the foot switch connected to X1 to cause the jar to move to position 2. At the same time a logic 1 will be inserted into M2 of shift register 1.
4. The jar will now be tracked through one complete index cycle using a 10-bit shift register, i.e. shift register 1, which uses M2 to M11.
5. At station 2, the presence of a signal in M4 will activate a pick and place robot system, which will pick up a lid and place it onto the jar.
6. At position 6, prior to where the screw lid is to be screwed on, i.e. when there is a signal in M6, a sensor connected to X3 is used to detect that the lid has been correctly placed on top of the jar or has not fallen OFF.
7. The lid is now tracked through the remainder of the cycle using a second shift register i.e. shift register 2, which uses M17 to M21.
8. At station 3, if there is a signal in M7, the jar shift register, and also in M17, the lid shift register, the lid screw ON cycle can be initiated.
9. Station 4 is the jar exit position and at position 10 if there is a signal in both M10 and M20, then a successful product has been produced and output Y4 will be energised to push the jar onto the accept conveyor. With a successful product being produced, i.e. a jar of hermetically sealed pickled onions, an accept display counter will be incremented.
10. If the robot did not place the lid correctly onto the top of the jar or it fell off between positions 4 and 6, then when that particular jar is rotated to position 11, output Y5 will be energised to push the jar onto the reject conveyor. With an unacceptable product being made, an unacceptable display counter will be incremented.

27.3 Shift register layout

Figure 27.2 shows the layout for the two shift registers, i.e. the index table shift register and the lid shift register and how they interface to the index table and the four stations.

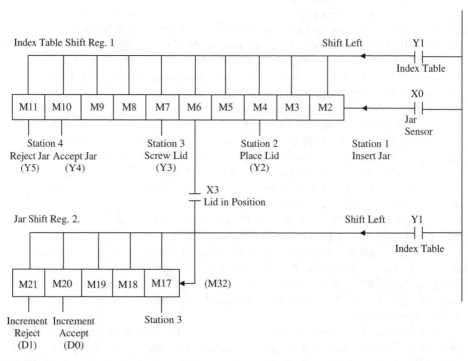

Figure 27.2

27.4 Ladder diagram – ROTARY1

```
         M8002
    0    ─┤├──┬─────────────────────────────[RST    D0    ]─
         Run │                                    Accept Total
         Pulse│
          X7 │
         ─┤├──┘─────────────────────────────[RST    D1    ]─
         Manual Reset                              Reject Total

* Station 1
          X0
    8    ─┤├───────────────────────────────────────(Y0    )─
         Jar Sensor                                GO Light

          X1       X2
   10    ─┤├──────┤├────────────────────────[PLS    M30   ]─
         Operator Index Complete                   Foot Switch
         Foot Sw. Sensor                           Pulse

          M30      T0
   14    ─┤├──┬──┤/├──────────────────────────────(Y1    )─
         Foot Sw │ Index Op                        Index
         Pulse   │ Time                            Rotate Sol
          Y1    │
         ─┤├───┤
         Index  │
         Rotate Sol
                │                                 [PLS    M31   ]─
                │                                  Timer
                │                                  Release
                │                                        K10
                └────────────────────────────────────(T0    )─
                                                   Index Op
                                                   Time

                                            *<Index Table Shift Reg1 M2-M11
          Y1
   25    ─┤├──────────────────────[SFTLP   X0    M2    K10   K1   ]─
         Index                            Jar
         Rotate Sol                       Sensor

* Station 2
          M4       T1
   35    ─┤├──────┤/├─────────────────────────────(Y2    )─
         Shift1   Robot Op                         Robot Pick/
         Pos 4    Time                             Place Lid

                   M31                                   K5
                  ─┤/├─────────────────────────────(T1    )─
                   Timer Release                   Robot Op
                                                   Time

                                            *<Jar Shift Reg2 M17-M21
          Y1
   44    ─┤├──────────────────────[SFTLP  M32   M17   K5    K1   ]─
         Index Rotate                     Shift2 Shift2
         Sol                              Input  Screw
                                          Signal Lid Pos 7
```

(*Continued*)

286 *Mitsubishi FX Programmable Logic Controllers*

```
          M6        X3
         ─┤├───────┤├───────────────────────────────────────(M32    )
  54    Shift1   Lid Detect                                 Shift 2
        Detect   Lid Sensor                                 Input Signal
        Position

* Station 3
          M7       M17        T2
         ─┤├──────┤├─────────┤/├──────────────────────────(Y3     )
  57    Shift1   Shift2    Screw Lid                        Screw
        Screw    Screw     ON Timer                         Lid ON
        Lid Pos7 Lid Pos7
                            M31                                      K5
                           ─┤/├──────────────────────────(T2     )
                           Timer                            Screw Lid
                           Release                          ON Timer

* Station 4
          M10      M20        T3
         ─┤├──────┤├─────────┤/├──────────────────────────(Y4     )
  67    Shift1   Shift2    Accept Jar                       Eject
        Pos 10   Pos 10    Timer                            Accept Jar

                                        ─────────────────[INCP   D0   ]
                                                            Accept
                                                            Total
                            M31                                      K5
                           ─┤/├──────────────────────────(T3     )
                           Timer                            Accept Jar
                           Release                          Timer

          M11      M21        T4
         ─┤├──────┤/├─────────┤/├──────────────────────────(Y5     )
  80    Shift1   Shift2    Reject Jar                       Eject
        Pos 11   Pos 11    Timer                            Reject Jar

                                        ─────────────────[INCP   D1   ]
                                                            Reject
                                                            Total
                            M31                                      K5
                           ─┤/├──────────────────────────(T4     )
                           Timer                            Reject Jar
                           Release                          Timer

         M8000
        ─┤├─────────────────────────────────────[BCD   D0   K4Y10 ]
  93    Run                                                Display Accept
                                                           Total

  99   ─────────────────────────────────────────────────────[END ]
```

27.5 Principle of operation – ROTARY1

1. Line 0
   (a) M8002 closes for one scan period, when the PLC is switched to Run.
   (b) In parallel with M8002 is a manual reset X7.
   (c) Hence when either the PLC is switched to Run or X7 is operated, data register D0 which is being used to totalise the number of acceptable products and data register D1 which is being used to totalise the number of rejected products, will be reset.
2. Line 8 **Station 1**
   (a) At position 1, a sensor X0 will give an output signal, when a jar enters the index table.
   (b) This will cause the Y0 light comes ON to indicate that a jar has been inserted into the rotary index table, where it is then filled with pickled onions.
3. Line 10
   (a) Mounted below the index table and rotating with it are 12 metal lugs. When the table has finished rotating and provided it has rotated the correct distance, then one of these lugs will be exactly opposite an inductive proximity switch X2. Hence an output from this switch signals that the table is exactly aligned for the jars to enter and exit the table.
   (b) The operator on seeing both the GO light and that the jar has been filled with pickles will operate a foot switch, which is connected to input X1.
   (c) The closure of the X1 contacts and the index complete X2 contacts will cause the M30 contacts to pulse, i.e. close for one scan time.
4. Line 14
   (a) The momentary closure of M30 will cause the Output Y1 to energise and latch.
   (b) Energising Y1 will cause the index rotary table to rotate, i.e. index 30° in an anti-clockwise direction.
   (c) The pulsing of M31 ensures that when there is a continuous supply of jars in the rotary table and hence all bits in both shift registers are always ON, the operations at stations 2, 3 and 4 will not stay latched ON.
   Timer T0 ensures that the index operation will be completed in exactly 1 sec.
5. Line 25
   (a) The closure of the Y1 contacts each time the index table rotates, sends a clock pulse signal to both shift register 1 and shift register 2.
   (b) Shift register 1 is produced using the instruction -[SFTLP X0 M2 K10 K1]-.
      i. SFTLP   Shift left one position each time the clock signal Y1 turns ON.
      ii. X0   The input bit data is provided by the logic state of the jar sensor X0.
      iii. M2   The 1st bit in the shift register is M2. Hence if X0 = 1, then on the first clock pulse, i.e. when the index table rotates for the first time, a logic 1 will be shifted into M2.
      iv. K10   This determines the number of bits or length of the shift register, i.e. shift register 1 = M2 – M11.
      v. K1   Each time when Y1 is energised and the index table rotates then the logic state of X0 is shifted into M2 and the existing contents of shift register 1 are shifted one place left.

6. Line 35   **Station 2 robot pick and place lid**
   (a) When the first jar reaches station 2, then there will be a logic 1 in bit M4 of shift register 1.
   (b) This will cause output Y2 to initiate the Robot pick and place operation.
   (c) The robot arm will pick up a lid and then place it on top of the jar of pickles.
   (d) The time for this operation being controlled by the 0.5-sec timer T1.
7. Line 44
   (a) As mentioned in Item 5, the closure of the Y1 contacts sends a clock pulse to both shift register 1 and shift register 2.
   (b) Shift register 2 is produced using the instruction -[SFTLP M32 M17 K5 K1]-.
      i. SFTLP   Shift left one position each time the clock signal Y1 turns ON.
      ii. M32   The input bit data is provided by the combined logic states of M6 from shift register 1 and the lid sensor X3 – Line 54. If both M6 and X3 are ON, then M32 will also be ON.
      iii. M17   The first bit in shift register 2 is M17. If M32 is ON, then on the next clock pulse, a Logic 1 will be shifted into M17.
      iv. K5   This determines the number of bits or length of shift register 2. shift register 2 = M17 – M21.
      v. K1   Each time when Y1 is energised and the index table rotates then if M32 is ON, a logic 1 is shifted into M17 and the existing contents of shift register 2 are shifted one place left.
8. Line 54
   (a) The shift 2 input signal M32 will turn ON, if both of the following signals are present.
      i. The M6 bit in shift register 1.
      ii. The lid detect sensor X3.
   (b) The reason for this line of the program occurring after the shift register 2 instruction SFTLP is described on page 291.
9. Line 57   **Station 3**
   (a) When the first jar reaches station 3, the M6 bit in shift register 1 will be ON and providing there is a lid correctly placed on top of the jar then the M17 bit of shift register 2 will also be ON.
   (b) These two signals will now activate the screw lid mechanism to screw the lid tightly onto the jar to ensure the pickles are hermetically sealed within the jar.
   (c) This operation is also carried out within 0.5 sec, as set by timer T2.
10. Line 67   **Station 4**
    (a) When the first acceptable jar of pickles reaches position 10, both the M10 bit of shift register 1 and the M20 bit of shift register 2 will be ON and hence this will enable the eject accept jar solenoid Y4 to be energised. This will cause the accept jar of pickles to be ejected from the rotary index table, onto the accept conveyor.
    (b) This operation is also carried out within 0.5 sec, as set by Timer T3.
    (c) Coincidently data register D0 will be incremented.
    (d) The function of D0 is to provide a count of the total number of acceptable products produced.

11. Line 80
    (a) If there is not a lid on the jar when it reaches position 11, then bit M11 of shift register 1 will be ON, but bit M21 of shift register 2 will be OFF and hence this will enable the eject reject jar solenoid Y5 to be energised. This will cause the unacceptable Jar of pickles to be ejected from the rotary index table, onto the reject conveyor.
    (b) This operation is also carried out within 0.5 sec, as set by timer T4.
    (c) Coincidently data register D1 will be incremented.
    (d) The function of D1 is to provide a count of the total number of rejected products. Obviously if this count reaches an unacceptable level, then steps would be taken to identify the problem and correct it.
12. Line 93
    The instruction -[BCD D0 K4Y10]- will enable the total number of acceptable jars of pickled onions produced, to be displayed on the seven-segment displays.

    The total shown on the seven-segment display provides the management with valuable information concerning the number of acceptable products made and the reliability of the production line producing them.

    For example if a required number are not produced in a given time due to breakdowns of the production line, then management has to decide on what level of support has to be given to improve its reliability.

27.6 Monitoring procedures

To simulate the operation of the program ROTARY1, press F3 to monitor the ladder diagram and then carry out the following:

Procedure no. 1

1. **Operate input X0**.
   This simulates the jar sensor.
2. **Operate and leave ON input X2**.
   This simulates the index complete sensor.
3. **Pulse input X1 ON and OFF**.
   This simulates the operator foot switch.
4. At line 25, M2 the first bit in shift register 1, will now be ON.
5. **Turn input X0 OFF**.
   This ensures only one jar will move round within the rotary table.
6. **Pulse input X1 twice**.
   Output Y2, which controls the operation of the pick and place robot, will now be energised for 0.5 sec. At line 35, bit M4 of shift register 1 will be ON.
7. **Pulse input X1 again twice**.
   At line 54, bit M6 of shift register 1 will now be ON.
8. **Operate input X3**.
   This simulates the presence of a lid at position 6. At lines 44 and 54, M32 will now be ON.

290 Mitsubishi FX Programmable Logic Controllers

9. **Pulse input X1 on just once.**
   Output Y3, which controls the operation of the lid screw cycle, will now be energised for 0.5 sec.
10. **Turn input X3 OFF.**
    As there have been no further jars inserted into the rotary table, no further lids will have been picked and placed onto the jars. Hence there will be no signal from the lid sensor.
11. **Pulse input X1 three times.**
    Note the following at line 67.
    (a) Output Y4, the eject accept jar solenoid will be energised.
    (b) Both outputs will be on for 0.5 sec.
    (c) Data register D0 will have been incremented to 1.
12. The seven-segment display will also be displaying a 1. This shows that the first jar of pickled onions has been successfully produced.

Procedure no. 2

For procedure no. 2, assume there is a continuous flow of jars.

1. **Operate and leave ON the following inputs.**
   (a) X0   Jar sensor.
   (b) X2   Index complete sensor.
   (c) X3   Lid detect sensor.
2. **Continuously pulse input X1.**
   This simulates the continuous operation of the foot switch.
3. Use device batch to monitor the flow of data through shift registers 1 and 2 (see Figure 27.3).

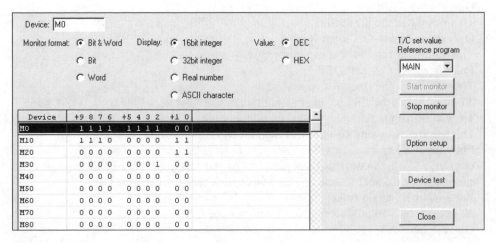

Figure 27.3

Procedure no. 3

For procedure no. 3 assume all of the jars are rejects in that none of the jars have had lids screwed onto them.

1. **Operate and leave ON just the following Inputs, i.e. leave Input X3 OFF**.
   (a) X0   Jar sensor.
   (b) X2   Index complete sensor.
2. **Continuously pulse input X1, the operator foot switch**.
3. Note as the jars reach position 11, the eject reject jar solenoid Y5 will now be energised each time input X1 is pulsed.

27.7 Instruction scan and execution

The program ROTARY1 is one of the rare occurrences in ladder diagram programming where for the program to work correctly, an instruction must be placed after another instruction, i.e. the output M32 must come *after* the shift register 2 instruction.

That is, in the order of instruction scanning and execution, the shift register 2 instruction will be executed before the output M32 instruction is executed.

To demonstrate this, carry out the following procedure:

1. Clear shift registers 1 and 2.
   (a) Monitor shift register 1, i.e. M2–M11 and shift register 2, i.e. M17–M21 as shown on page 290.
   (b) Operate just input X2, the index complete sensor.
   (c) Pulse input X1, the operator foot switch until both shift registers are cleared.
2. Let just one jar enter the index table
   (a) Operate input X0, the jar sensor.
   (b) Operate and leave ON input X2, the index complete sensor.
   (c) Operate and leave ON input X3, the lid sensor.
   (d) Pulse input X1 just once.
   (e) Turn input X0 OFF, this ensures no more jars will enter the index table.
3. Pulse input X1, until M6 = 1.
4. Consider the shift register layout drawing on page 284.
   (a) After the fifth clock pulse, M6 = 1, M5 = 0 and M32 = 1.
   (b) Hence with M32 ON, the data input to shift register 2 = 1.
5. After the sixth clock pulse, M7 = 1, M6 = 0, M17 = 1 and M32 = 0. However, since the output M32 instruction is after the shift 2 instruction, then M5 and hence M32 will still be ON, at the instant when the sixth clock pulse is applied to shift register 2.
6. The logic state of M32 will not change to OFF until the output M32 instruction at line 54 is executed, which is after the shift 2 instruction.

7. Therefore since M32 is still ON, when the sixth clock signal is applied to shift register 2, a logic 1 will be clocked into bit M17 of shift register 2.
8. Had the output M32 instruction occurred before the shift register 2 instruction then when the sixth clock pulse is applied to shift register 2, M32 would be OFF.

    Therefore although a lid had been detected, a logic 0 would have been clocked into bit M17 of shift register 2 instead of a logic 1.

28

Index registers V and Z

Index registers are similar to data registers in that they can store 16-bit data. However, the main function of these registers is to provide an offset value to other PLC devices.

In all of the FX range of PLCs, there are a total of 16 index registers.
V0–V7;
Z0–Z7.

28.1 Index register instructions

The following is a list of Index register instructions used in this chapter.

(a) **MOV K5 Z0** Load index register Z0 register with the constant value of 5.
(b) **MOV D20Z0 D1** With Z0 = 5, move the contents of D(20 + 5) into D1, i.e. move the contents of D25 into data register D1.
(c) **MOVP D2 Z0** Move on a closing input, the contents of D2 into index register Z0.
(d) **INCP V0** Increment on a closing input, the contents of index register V0.
(e) **DIVP V0 D0 D1** Divide the contents of index register V0 by the contents of D0 and store the results, i.e. 11/4, as follows:
  D1 contains the quotient = 2.
  D2 contains the remainder = 3.
(f) **BCD D20Z0 K2Y10** If Z0 = 0, output the first 8 bit contents of D20 to Y10–Y17 in BCD, i.e. display the contents of D20 on the seven-segment displays.
  If Z0 = 1, display the contents of D21.
  ⋮
  If Z0 = 5, display the contents of D25.
(g) **BIN K2X10 D20Z0** Input in BCD format the contents of the thumbwheel switches which are connected to inputs X10–X17 and then transfer those contents to a look-up table whose base address is D20.
  The actual data register in which the thumbwheel contents will be stored will depend on the value in the Z register.

28.2 Stock control application – INDEX1

The following application is part of a system, in which the contents of four data registers D20–D23 can be entered into a look-up table and then displayed.

This could be used in a stock control system, in which four different components are constantly being put into and removed from a storage area.

In practice a bar-code reader would be used to read the type and number of components coming into the warehouse and then using a PLC, transfer the data to the appropriate data register in the look-up table.

Such a system would enable constant monitoring of the stock levels of the four components.

28.3 System block diagram

Figure 28.1 shows the input/output signals for the system.

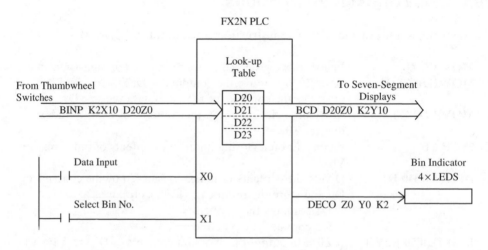

Figure 28.1

28.4 Warehouse – look-up table

Data Register	Contents	Bin No.
D20	40	1
D21	85	2
D22	70	3
D23	15	4

28.5 Ladder diagram – INDEX1

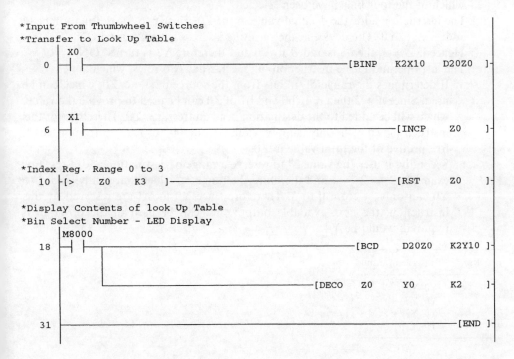

28.6 Principle of operation

1. Line 0
   (a) The operation of the start button X0, enables the instruction -[BINP K2X10 D20Z0]- to be executed.
   The thumbwheel input data, which is in the range 0–99, will now be transferred to the look-up table D20 to D23 each time input X0 is pulsed ON/OFF.
   (b) The register that is selected to store the data will depend upon the value in the index register Z0, i.e. 0–3.
   For example, if Z0 = 0, then the contents of the thumbwheel switches will be transferred to D20 in the look-up table, when input X0 is pulsed.
2. Line 6
   Once the data has been stored into the look-up table, Index register Z0 is incremented by operating input X1.
3. Line 10
   The instructions -[> Z0 K3]- and -[RST Z0]- ensure that the value in the Z register is always in the range 0–3.
4. Line 18
   (a) The instruction -[BCD D20Z0 K2Y10]- enables the value stored in the selected register of the look-up table, i.e. data registers D0–D3, to be displayed on the seven-segment displays.

(b) The instruction -[DECO Z0 Y0 K2]- is used to display on the LEDs Y0–Y3, which of the four bins have been selected.
(c) The instruction adds the decimal value in the Z0 register to the destination head address, i.e. Y0. The device at the resulting address is then turned ON.
(d) Hence if Z = 3, then 3 is added to Y0 and therefore Y3 is turned ON.
(e) The last operand in the instruction 'n', i.e. K2 has two roles, which are:
  i. It determines the range of 'n' bits from the source data, i.e. Z0, which can be used. Since n = 2 then only b0 and b1 of Z0 can be used to provide the value, which will be added to the destination head address, i.e. Y0. Therefore for this instruction Z0 will only supply values from 00 to 11, i.e. from 0 to 3, irrespective of the total value in Z0.
  ii. Secondly, it uses the value 2^n to reserve a range of destination addresses. For example, $2^n = 2^2 = 4$ and therefore the range of values that can be added to Y0 can only be from 0 to 3. Therefore since Y0 to Y3 are reserved by this instruction, the next available output that can be used in the program if required, would be Y4.

n	Range of reserved outputs
1	Y0–Y1
2	Y0–Y3
3	Y0–Y7
4	Y0–Y15

28.7 Monitoring – INDEX1

1. Enter the following data numbers in turn from the thumbwheel switches and after each number has been entered, pulse input X0 to enable them to be stored into the look-up table.

    D20   40
    D21   85
    D22   70
    D23   15

2. Monitor the following as shown in Figure 28.2, using entry data monitor

Device	ON/OFF/Current	Setting value	Connect	Coil	Device comment
Z0	3				
D20	40				
D21	85				
D22	70				
D23	15				
K2Y010	21				

Figure 28.2

3. Note: With index register Z0 = 3, the contents of D23, i.e. 15 are output in BCD format to the seven-segment displays, which are connected to Y10–Y27.

D23	BCD output	Equivalent decimal output (K2Y10)
15	0001 0101	21

29

Recipe application – BREW1

This project requires that five different beers be produced, by selecting the correct proportions from three basic ingredients. This enables the correct recipe to be obtained for the selected beer.

Let the beers be:

(a) Beer 1
(b) Beer 2
(c) Beer 3
(d) Beer 4
(e) Beer 5.

Let the ingredients be:

(a) Hops
(b) Wheat
(c) Barley.

This project basically consists of two separate areas:

(a) The recipe data
(b) The ladder diagram.

As will be shown starting on page 300, the recipe data can be entered into Gx-Developer and then downloaded on its own into the PLC. After which the ladder diagram can then be entered and downloaded.

29.1 System diagram

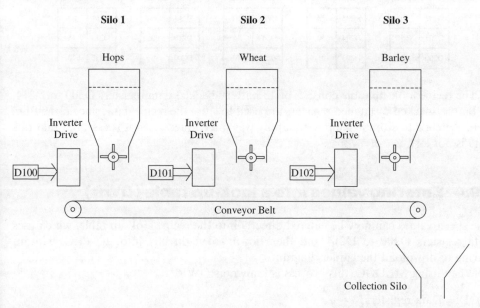

Figure 29.1

29.2 Sequence of operation

1. Depending on the type of beer, which is to be brewed, different proportions of hops, wheat and barley have to be used (see Figure 29.1).
2. The required proportions are taken from a recipe look-up table and transferred to data registers D100, D102, D103.
3. The values in the three data registers are then sent to three separate inverter drive units, which control the speed of three motors.
4. As each motor rotates at different speeds, different amounts of hops, wheat and barley will be delivered onto the conveyor and then into a collection silo.
5. The different amounts of the three products will then be used to brew the selected beer.

29.3 Recipe look-up tables

Each separate beer, takes three speed values from a 15-value recipe look-up table. These three values are the speeds at which the three Silo motors rotate at to produce the correct proportions of the three ingredients, for the beer about to be brewed.

300 *Mitsubishi FX Programmable Logic Controllers*

Beer 1		Beer 2		Beer 3		Beer 4		Beer 5	
Reg.	Speed	Reg.	Speed	Reg.	Speed	Reg.	Speed	Reg.	Speed
D202	20	D205	35	D208	50	D211	65	D214	80
D201	15	D204	30	D207	45	D210	60	D213	75
D200	10	D203	25	D206	40	D209	55	D212	70

The recipe look-up table consists of 15 battery-backed data registers D200 to D214.

Battery-backed or latched registers are used to store the recipe data, since if the PLC were switched to stop or the power supply was interrupted, then it is essential that this data is not lost.

29.4 Entering values into a look-up table (DWR)

The speed values can now be entered directly into the recipe look-up table, which uses data registers D200 to D214 and then downloaded directly into the PLC, without having to download the ladder diagram.

When using MEDOC, this process is known as DWR.

(a) D = Data registers
(b) W = Network registers
(c) R = File registers.

1. Enter a new project BREW1.
2. Select the project data list as described on pages 22 and 146.
3. From the project data list, select device memory (see Figure 29.2).

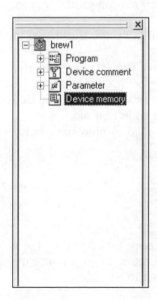

Figure 29.2

4. Use the right-hand mouse button to obtain the display shown in Figure 29.3.

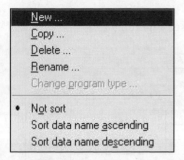

Figure 29.3

5. Select New. By default the name given to the device memory data will be MAIN (see Figure 29.4).

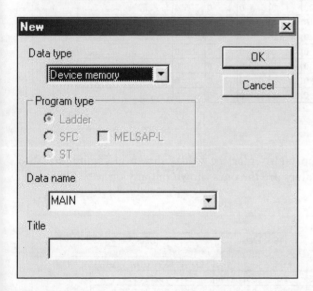

Figure 29.4

6. Select OK to obtain the display shown in Figure 29.5.

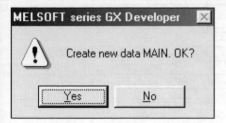

Figure 29.5

302 *Mitsubishi FX Programmable Logic Controllers*

7. Select Yes, to create a new data file for storing device memory data with the file name MAIN.
8. The device memory display now appears.
9. Change the device label to D200 and select display (see Figure 29.6).

Device Label	D200			Display		16-bit integer			DEC		D0–D7999
Device name	0	1	2	3	4	5	6	7	Character string		
D200	0	0	0	0	0	0	0	0	....		
D208	0	0	0	0	0	0	0	0	....		
D216	0	0	0	0	0	0	0	0	....		
D224	0	0	0	0	0	0	0	0	....		
D232	0	0	0	0	0	0	0	0	....		
D240	0	0	0	0	0	0	0	0	....		
D248	0	0	0	0	0	0	0	0	....		
D256	0	0	0	0	0	0	0	0	....		
D264	0	0	0	0	0	0	0	0	....		
D272	0	0	0	0	0	0	0	0	....		
D280	0	0	0	0	0	0	0	0	....		
D288	0	0	0	0	0	0	0	0	....		
D296	0	0	0	0	0	0	0	0	....		
D304	0	0	0	0	0	0	0	0	....		

Figure 29.6

10. To produce the recipe look-up table, enter the values shown in Figure 29.7 into D200 to D214. These speed values are the ones shown initially on page 300.

Device Label	D200			Display		16-bit integer			DEC		D0–D7999
Device name	0	1	2	3	4	5	6	7	Character string		
D200	10	15	20	25	30	35	40	45	..........#.(.-.		
D208	50	55	60	65	70	75	80	0	2.7.<.A.F.K.P...		
D216	0	0	0	0	0	0	0	0	....		
D224	0	0	0	0	0	0	0	0	....		
D232	0	0	0	0	0	0	0	0	....		
D240	0	0	0	0	0	0	0	0	....		
D248	0	0	0	0	0	0	0	0	....		
D256	0	0	0	0	0	0	0	0	....		
D264	0	0	0	0	0	0	0	0	....		
D272	0	0	0	0	0	0	0	0	....		
D280	0	0	0	0	0	0	0	0	....		
D288	0	0	0	0	0	0	0	0	....		
D296	0	0	0	0	0	0	0	0	....		
D304	0	0	0	0	0	0	0	0	....		

Figure 29.7

29.5 Downloading the recipe look-up table

The recipe look-up table will now be downloaded directly into the PLC.

1. Select Online.
2. Select Write to PLC.
3. Highlight Parameter and Device memory MAIN (see Figure 29.8).

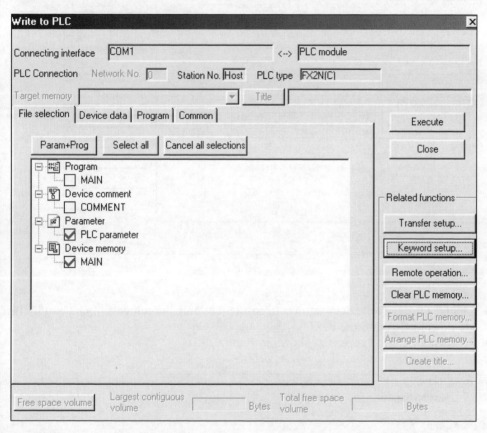

Figure 29.8

29.6 Selecting the device memory range

Instead of downloading into the PLC all of the device memory values, it is possible to select just those devices and the range of addresses, which are required. For this project, it is only necessary to download data registers D200–D214.

1. From the Write to PLC display select Device data.
2. Cursor down until the Default Data Register download range is obtained.
3. The display should appear as shown in Figure 29.9.

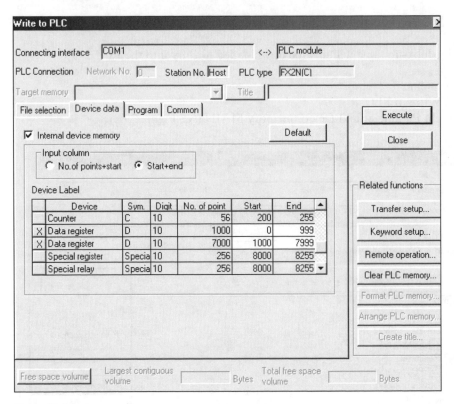

Figure 29.9

4. Change the device memory settings to ensure that only the contents of D200 to D214 are downloaded into the PLC.
5. The display should appear as shown in Figure 29.10.

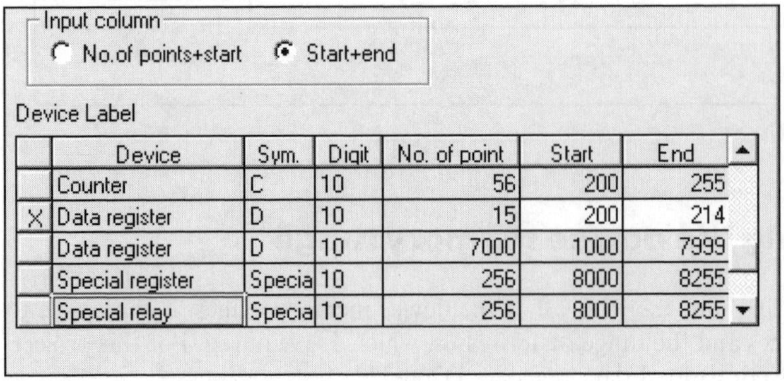

Figure 29.10

6. Select Execute to enable the parameters and the selected data register values to be downloaded to the PLC.

29.7 Monitoring the recipe look-up table values

Use device batch to monitor the recipe look-up table values and hence confirm they have been downloaded into the PLC (see Figure 29.11).

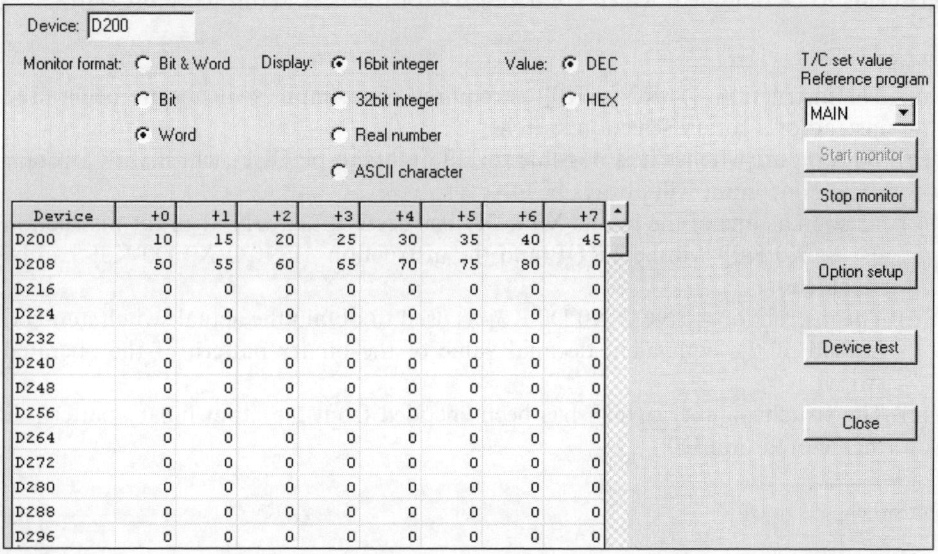

Device	+0	+1	+2	+3	+4	+5	+6	+7
D200	10	15	20	25	30	35	40	45
D208	50	55	60	65	70	75	80	0
D216	0	0	0	0	0	0	0	0
D224	0	0	0	0	0	0	0	0
D232	0	0	0	0	0	0	0	0
D240	0	0	0	0	0	0	0	0
D248	0	0	0	0	0	0	0	0
D256	0	0	0	0	0	0	0	0
D264	0	0	0	0	0	0	0	0
D272	0	0	0	0	0	0	0	0
D280	0	0	0	0	0	0	0	0
D288	0	0	0	0	0	0	0	0
D296	0	0	0	0	0	0	0	0

Figure 29.11

29.8 Ladder diagram – BREW1

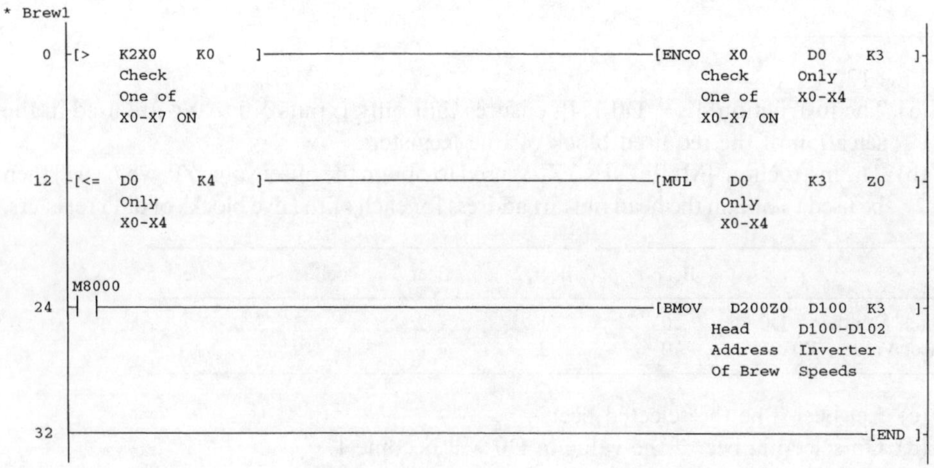

```
* Brew1
    0  ─[> K2X0  K0 ]─────────────────────────[ENCO  X0    D0   K3 ]─
           Check                                     Check Only
           One of                                    One of X0-X4
           X0-X7 ON                                  X0-X7 ON

   12  ─[<= D0   K4 ]─────────────────────────[MUL   D0    K3   Z0 ]─
           Only                                      Only
           X0-X4                                     X0-X4

        M8000
   24  ──┤ ├─────────────────────────────────[BMOV  D200Z0 D100 K3 ]─
                                                    Head   D100-D102
                                                    Address Inverter
                                                    Of Brew Speeds

   32  ──────────────────────────────────────────────────────[END ]─
```

Note: The Ladder diagram BREW1 will now have to be downloaded to the PLC since as yet, only the parameters and the device memory values D200–D214, have been downloaded.

29.9 Principle of operation – BREW1

The function of the ladder diagram is to enable a particular beer to be selected and from the selection obtain the required motor speeds, which will then be sent to the 3 × inverter drives.

The different motor speeds enable the correct proportions of the three different ingredients to be obtained, which are necessary for the beer about to be brewed.

1. Line 0
   (a) The instruction -[> K2X0 K0]- is required, since input switches are being used instead of a rotary selection switch.
   (b) With input switches it is possible for all inputs to be OFF, whilst with a rotary switch one input will always be ON.
   (c) As soon as one of the inputs X0 to X7 has been turned ON, then the instruction -[> K2X0 K0]- will be TRUE and the instruction -[ENCO X0 D0 K3]- can be executed.
   (d) The instruction -[ENCO X0 D0 K3]- is used to obtain the actual switch number, instead of the equivalent decimal value of the binary pattern of the operated switch.
   (e) The switch number, which has been encoded from the input binary pattern, is then stored into D0.

Input switch	D0
X0	0
X1	1
X2	2
X3	3
X4	4
X5	5
X6	6
X7	7

2. Line 12
   (a) The instruction -[<= D0 K4]- ensures that only Inputs X0 to X4 are used in the selection of the required block of File Registers.
   (b) The instruction -[MUL D0 K3 Z]- is used to obtain the offset value Z0, which will then be used to obtain the head or start address for each of the five blocks of data registers.

	Beer 1	Beer 2	Beer 3	Beer 4	Beer 5
Switch Number – D0	0	1	2	3	4
Offset Value – Z0	0	3	6	9	12

   (c) Let beer 2 be the selected beer.
   (d) On selecting beer 2 the value in D0 will become 1.
   (e) This value is then multiplied by 3 and stored into the index register Z0. Hence Z0 will contain 3.
   (f) The value stored in index register Z0 is the selected switch number × 3.

3. Line 24
   (a) The instruction -[BMOV D200Z0 D100 K3]- is used to transfer a block of three speed values from the recipe table (D200–D214) to the inverter drives via D100, D101 and D102.
   (b) Index register Z0 contains the offset value, which is automatically added to the base address of the recipe table, i.e. D200 to obtain the head address of the required block of data registers.

	\multicolumn{5}{c}{Switch Number – D0}
	0	1	2	3	4
Selected Beer	Beer 1	Beer 2	Beer 3	Beer 4	Beer 5
Head Address D200Z0	D200	D203	D206	D209	D212

   (c) The K3 part of the instruction, causes three moves of the selected recipe data to the inverter drive speed registers, D100, D101 and D102.

	Switch No. 0	Switch No. 1	Switch No. 2	Switch No. 3	Switch No. 4
	Beer 1	Beer 2	Beer 3	Beer 4	Beer 5
Data Regs	D200 D201 D202	D203 D204 D205	D206 D207 D208	D209 D210 D211	D212 D213 D214
Inv. Regs	D100 D101 D102	D100 D101 D102	D100 D101 D102	D100 D101 D102	D100 D101 D102

   (d) Hence on selecting beer 2, by operating input X1, the following transfers will take place.

   Recipe data transfer for beer 2

Source Recipe Data Registers	Contents Speed – RPM	Ingredients	Destination Silo Inverter Drive Registers
D203	25	Hops	D100
D204	30	Wheat	D101
D205	35	Barley	D102

4. On selecting beer 2, the silo motors will rotate at the above speeds, which will enable the correct proportions of the three ingredients to be transferred onto the conveyor belt and delivered to the collection silo.
5. The brewing process for beer 2 can then take place.

29.10 Monitoring – BREW1

1. Use entry data monitoring, to monitor the following:
   (a) X0–X4   Beer selection switches.
   (b) Z0   Index Register.

(c) D0   Beer selection number.
(d) D100 Hops motor speed.
(e) D101 Wheat motor speed.
(f) D102 Barley motor speed.

2. From the ladder diagram, press F3 for monitor.
3. Press the right-hand mouse button to obtain the display shown in Figure 29.12.

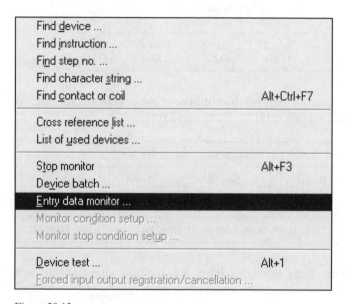

Figure 29.12

4. Register the devices as shown in Figure 29.13 and then monitor them.

Device	ON/OFF/Current	Setting value	Connect	Coil	Device comment
X000			0		Check One of X0-X7 ON
X001			1		
X002			0		
X003			0		
X004			0		
D0		1			Only X0 - X4
Z0		3			
D100		25			D100-102InverterSpeeds
D101		30			
D102		35			

Figure 29.13

29.11 Test results

By operating each of the switches X0 to X4 in turn, the following results should be obtained.

Input Switch	Selected Brew	Inverter Drive Speeds		
		Hops (D100)	Wheat (D102)	Barley (D103)
X0	Beer 1	10	15	20
X1	Beer 2	25	30	35
X2	Beer 3	40	45	50
X3	Beer 4	55	60	65
X4	Beer 5	70	75	80

29.12 Excel spreadsheet – recipe1

Using Microsoft Excel, the recipe look-up table can be entered and saved as an Excel spreadsheet file, i.e. recipe1 (see Figure 29.14).

Then using copy and paste it is possible to transfer the recipe look-up table from the Excel Spreadsheet to the device memory in Gx-Developer for the program BREW1.

This procedure enables the recipe data to be easily changed, whenever necessary.

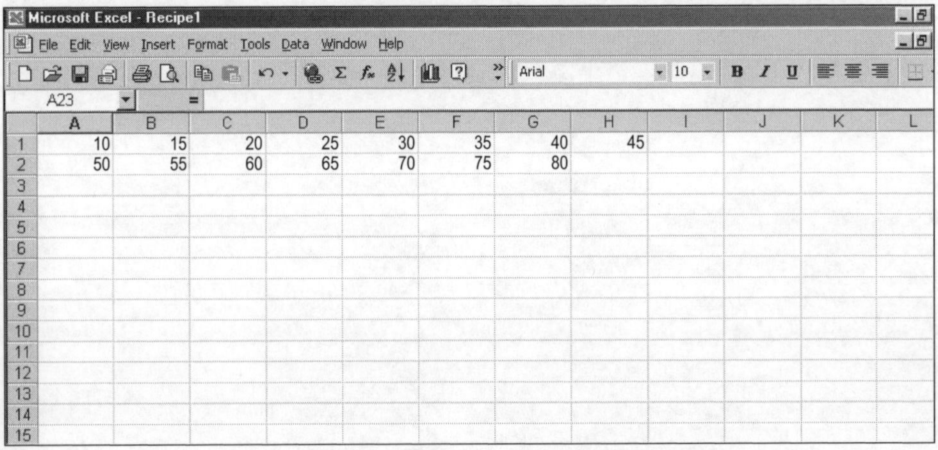

Figure 29.14

30

Sub-routines

A sub-routine is a separate section of a ladder diagram, which carries out a specific task and which is usually placed after the main section.

The sub-routine is called from a known part of the main program using the instruction CALL.

Once the sub-routine instructions have been executed then the execution of the instruction SRET causes the program to return to the main section and execute the instructions immediately following the CALL instruction.

The diagram shown on page 311 shows that the main program consists simply of three Calls to three different Sub-routines.

30.1 Sub-routine program flow

Main program

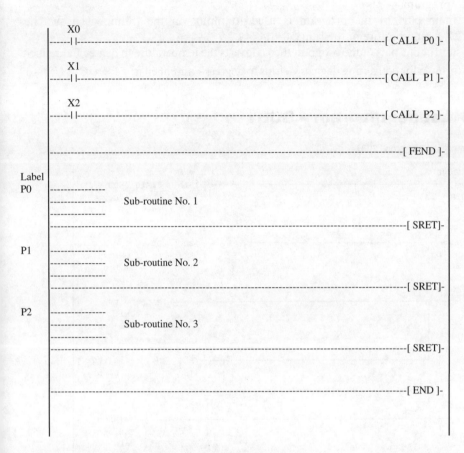

30.2 Principle of operation

1. When input X0 is ON, the execution of the CALL instruction will cause a jump to the sub-routine instructions, which starts at label P0.
   Note: In the FX2N range of PLCs a total of 63 labels are available, in the range P0–P62.
2. The sub-routine instructions are now executed until the sub-routine return -[SRET]- instruction is executed.
3. The program will now return to the main program, to enable the remaining main program instructions to be executed, or as shown on the above diagram to CALL the next sub-routine.
4. Note: At the end of the main program instructions there must be a FEND – First End Instruction – to designate the end of the main program. This ensures that only the main program instructions will be scanned and executed, when there is no CALL to a sub-routine.

30.3 Temperature conversion – SUB1

The program SUB1 is used to demonstrate the use of a sub-routine within a ladder diagram program.

The main part of the program is used to input via the thumbwheel switches X10–X27, a temperature in the range 0–100 °C.

The sub-routine part of the program then converts the temperature to degree Fahrenheit.

The formula for converting degrees Celsius to degrees Fahrenheit is: °F = (°C × 9/5) + 32

30.4 Ladder diagram – SUB1

```
*Sub-Routine Program-SUB1
*Main Program
       |M8000
    0  | |————————————————————————————————[BIN    K3X10    D0    ]
       Run                                         Input    Deg.C
                                                   Deg.C

       X0
    6  |/|————————————————————————————————[BCD    D0       K4Y10 ]
       OFF Dg C                                    Deg.C    Display
       ON  Dg F                                             Deg.C

       X0
   12  | |————————————————————————————————[CALL            P0    ]
                                                            Call
                                                            Sub-Routine

        ————————————————————————————————[BCD     D6       K4Y10 ]
                                                   Display
                                                   Deg. F

   21  ——————————————————————————————————[FEND                   ]
                                                   End of
                                                   Main Prog.
*Sub-Routine
P0     |M8000
   22  | |————————————————————————————————[MUL    D0    K9    D2 ]
       Run                                         Deg.C        Mulx9

        ————————————————————————————————[DIV     D2    K5    D4 ]
                                                               Div/5

        ————————————————————————————————[ADD     D4    K32   D6 ]
                                                               Add+32

   45  ——————————————————————————————————[SRET                   ]
                                                   Return to
                                                   Main Prog.
                                                   D6=Deg. F

   46  ——————————————————————————————————[END                    ]
```

30.5 Labels

Labels are in the range from P0 to P62.

A label is entered at the point on the ladder diagram by moving the cursor to the left of the ladder diagram and then enter into the required label number, i.e. P0.

30.6 Principle of operation – SUB1

1. Line 0
   The input temperature in degrees Celsius is entered via the thumbwheel switches X10–X27 and stored in data register D0.
2. Line 6
   With input X0 OFF, the instruction -[BCD D0 K4Y10]- enables the temperature in degrees Celsius to be displayed on the seven-segment displays Y10–Y27.
3. Line 12
   The operation of input X0 enables the following:
   (a) The call to the sub-routine at label P0.
   (b) The sub-routine enables the input temperature to be converted from degrees Celsius to degrees Fahrenheit.
   (c) The instruction -[BCD D6 K4Y10]- will enable the return value from the sub-routine, i.e. the temperature in degrees Fahrenheit, to be displayed.

30.7 The sub-routine instructions

1. Line 22
   (a) This is the start of the sub-routine section of the program at Label P0.
   (b) The instruction -[MUL D0 K9 D2]- multiplies °C with 9 and stores the result in data registers D2 and D3.
   (c) The instruction -[DIV D2 K5 D4]- divides the contents of D2 by 5 and the result is stored in data registers D4 and D5.
   (d) Finally the instruction -[ADD DD K5 D4]- adds the constant value of 32 to the contents of D4 and the result is stored in data register D6.
   (e) Data register D6 now contains the converted value in degree Fahrenheit.
2. Line 45
   The instruction SRET returns the program back to line 12 of the main program, i.e. to the instruction -[BCD D6 K4Y10]-, which follows the CALL instruction.

30.8 Monitoring – SUB1

Monitor D0–D7, using device batch.

Figure 30.1 shows the contents of the data registers, when 85 °C is converted to 185 °F.

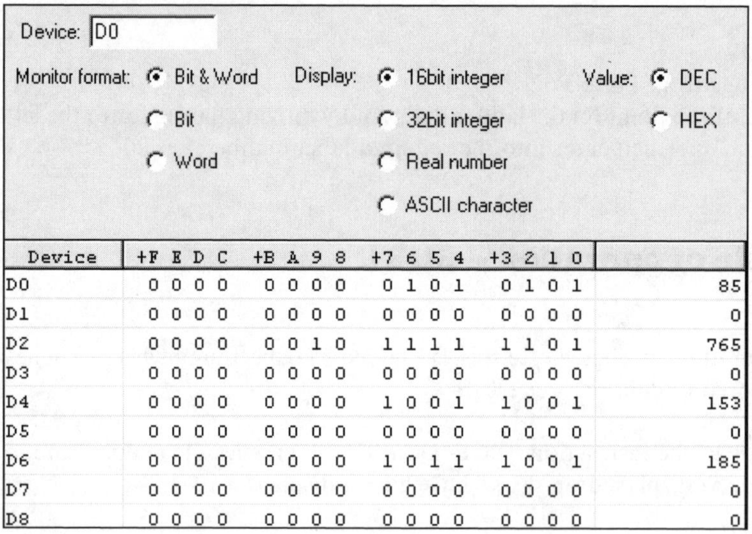

```
Device: D0
Monitor format:  ⦿ Bit & Word    Display:  ⦿ 16bit integer    Value:  ⦿ DEC
                 ○ Bit                     ○ 32bit integer            ○ HEX
                 ○ Word                    ○ Real number
                                           ○ ASCII character
```

Device	+F E D C	+B A 9 8	+7 6 5 4	+3 2 1 0	
D0	0 0 0 0	0 0 0 0	0 1 0 1	0 1 0 1	85
D1	0 0 0 0	0 0 0 0	0 0 0 0	0 0 0 0	0
D2	0 0 0 0	0 0 1 0	1 1 1 1	1 1 0 1	765
D3	0 0 0 0	0 0 0 0	0 0 0 0	0 0 0 0	0
D4	0 0 0 0	0 0 0 0	1 0 0 1	1 0 0 1	153
D5	0 0 0 0	0 0 0 0	0 0 0 0	0 0 0 0	0
D6	0 0 0 0	0 0 0 0	1 0 1 1	1 0 0 1	185
D7	0 0 0 0	0 0 0 0	0 0 0 0	0 0 0 0	0
D8	0 0 0 0	0 0 0 0	0 0 0 0	0 0 0 0	0

Figure 30.1

Check the temperatures shown below.

°C – D0	°F – D6
0	32
10	50
28	82
38	100
85	185
100	212

31

Interrupts

An interrupt is an input signal, which halts the execution of the main program and then immediately jumps to an interrupt service routine (ISR).

The ISR carries out a specific task and then returns back to the main program.

However, the main difference between an interrupt and a sub-routine CALL is that an interrupt can occur at any time during the execution of the main program, whereas a sub-routine CALL can only be carried out when, during the scan time of the main program, the CALL Instruction is executed.

If it is necessary to take immediate action in the event of a particular input signal being operated, then the use of an interrupt should be considered.

However, if immediate action were required in the event of an emergency stop situation, then a safety relay must be used and not the operation of an interrupt input.

31.1 Interrupt application

The following application is an example of how an interrupt can be used, with a sequential PLC control program, i.e. PNEU1 (refer Figure 8.2, page 92).

The interrupt signal in this program is the operation of Input X0.

The basis of this interrupt program is that the PLC is connected to a data logger computer, and at regular intervals, the data logger generates an interrupt signal by sending a +24 V signal to Input X0 of the PLC.

On receiving the interrupt signal, the main program completes the instruction currently being executed and then the PLC jumps to the ISR to execute the ISR instructions. These instructions require the number of complete machine cycles which have occurred, since the previous interrupt signal was generated to be output on Y10–Y17 in BCD format to the data logger computer.

In practice, the number of machine cycles would be output in serial format to the data logger.

The interrupt operation has no effect on the main program, i.e. the operation of Piston A and Piston B.

The data logger, on receiving the number of products made, can then produce the following management information:

1. The number of components or products produced in a given time.
2. The reliability of the machine, i.e. has any production time been lost due to breakdowns.

3. The total operating time of the machine, and from this information, produce Planned Preventive Maintenance schedules (PPMs).

Once the ISR instructions have been executed, program control returns back to the main program, and the PLC will start executing the main program instructions again, starting with the instruction following the one which was being executed when the interrupt occurred.

31.2 Interrupt project – INT1

The project INT1 simulates the same sequence of operations as PNEU1, but includes the following modifications:

1. Operates the Outputs Y0 and Y1 automatically in sequence, without the need for the simulated limit switches X2–X4.
2. An interrupt facility, which displays the number of completed cycles.

Basic system

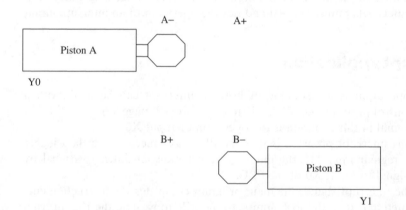

Figure 31.1

31.3 Sequence of operation – automatic cycle

The sequence of operations for the two pistons is as follows:

1. Start.
2. A+   Piston A OUT.
3. B+   Piston B OUT.
4. A–   Piston A IN.
5. 5-sec time delay.
6. B–   Piston B IN.
7. Repeat sequence   Item 2.

31.4 Waveforms

To determine the ON/OFF times for the Outputs Y0 and Y1, which control Pistons A and B respectively, the following waveforms were produced:

Piston wavefoms

The waveforms in Figure 31.2 show that the time taken for each piston to move OUT is 1 sec and then to move back IN is also 1 sec.

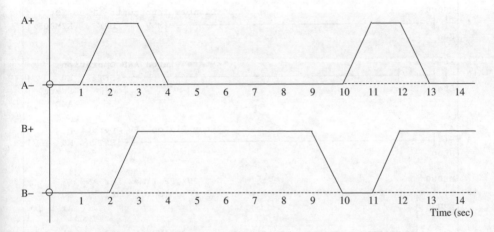

Output waveforms Y0 and Y1

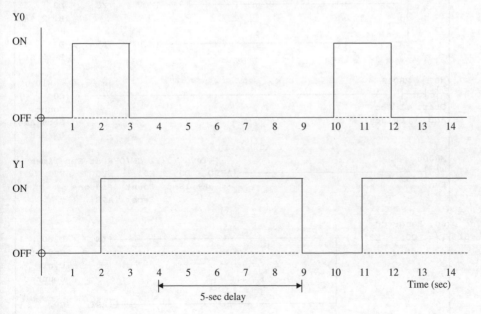

Figure 31.2

318 *Mitsubishi FX Programmable Logic Controllers*

31.5 Ladder diagram – INT1

Main Program

```
                                          *<Enable Interrupt Inputs
 0  ─────────────────────────────────────────────────────────[EI      ]─

    M8002                                 *<Enable X0 Interrupt
 1  ──┤ ├──────────────────────────────────────[RST    M8050  ]─
    Run                                                0-INT ON
    Pulse                                              1-INT OFF

    M8000                                 *<Disable Interrupts I1 to I8
 4  ──┤ ├──────────────────────────────[MOV    HFF    K2M8051]─
    Run

    X6    X7                              *<Start Piston A&B Operation
10  ──┤ ├──┤/├──────────────────────────────────[SET    M0     ]─
    Start Stop                                          Start
                                                        A&B

    X7                                    *<Stop Piston A&B Operation
13  ──┤ ├──────────────────────────────────────[RST    M0     ]─
    Stop

    M8000                                 Set ON/OFF Times for A&B
15  ──┤ ├───────────────────────────[MOV    K1     D0     ]─
    Run                                                 A+ 1 Sec

                                    ───────[MOV    K3     D1     ]─
                                                        A- 3 Sec

                                    ───────[MOV    K2     D2     ]─
                                                        B+ 2 Sec

                                    ───────[MOV    K9     D3     ]─
                                                        B- 9 Sec

    M0    M8013                                              K10
36  ──┤ ├──┤ ├──────────────────────────────────────────(C0      )─
    Start 1 Sec                                         Count
    A&B   Pulses                                        Secs A&B
                                                        ON/OFF

    M8000                                 *<Turn Y0/Y1 ON/OFF at Set Times
41  ──┤ ├────────────────[ABSD    D0    C0     Y0    K2   ]─
    Run                          A+ 1Sec Count  Pistons
                                         Secs  A&B
*C0 = 9
*Inc D200, Reset C0
51  ─[>=  C0    K9   ]──────────────────────────[INCP   D200   ]─
                                                        End of
                                                        Cycle
                                                        Counter

                                            ─────[RST    C0     ]─
```

Interrupts 319

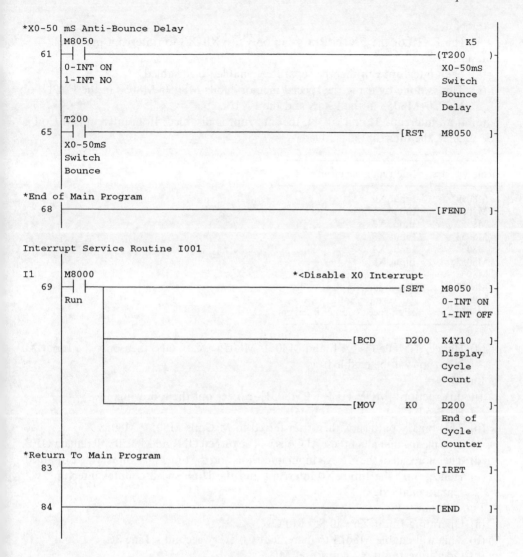

31.6 Principle of operation – INT1

Main program

The main program operates continuously as described in Section 31.3.

1. Line 0
    (a) The Instruction -[EI]- enables all the interrupts.
    (b) These are:
        (i) The X0–X5 Input interrupts.
        (ii) Timed interrupts.
        (iii) High-speed counter interrupts.

320 *Mitsubishi FX Programmable Logic Controllers*

   (c) Note:
   The FX1S or the FX1N PLCs can only use X0–X3 for interrupt purposes.
2. Line 1
   (a) Each interrupt can then be separately enabled or disabled.
   (b) This is done by using the special memory coils M8050–M8058 in the FX2N, or M8050–M8055 in the FX1S and the FX1N.
   (c) If an interrupt M coil is ON, the interrupt is disabled. If an interrupt M Coil is OFF, the interrupt is enabled.

Memory coil	Interrupt assignment
M8050	Input X0
M8051	Input X1
M8052	Input X2
M8053	Input X3
M8054	Input X4
M8055	Input X5
M8056	Timed interrupt 1
M8057	Timed interrupt 2
M8058	Timed interrupt 3
M8059	High-speed counter interrupt

   (d) Hence, if M8050 is OFF and M8051–M8059 are all ON, then only the Input X0 interrupt will be enabled.
3. Line 4
   The Instruction -[MOV HFF K2M8051]- carries out the following:
   (a) HFF = 1 1 1 1 1 1 1 1 in binary.
   (b) This binary pattern is then transferred to M Coils M8051–M8058.
   (c) This means that all of these M Coils will be turned ON and M8050 will remain OFF.
   (d) Therefore, all of the X1–X5 Input interrupts and the timer interrupts will be disabled.
   (e) Hence, only the Input X0 interrupt and the High-speed counter interrupt, will remain enabled.
4. Lines 10 and 13
   (a) Operating Input X6 will Set M0 ON.
   (b) This will enable M8013 to increment C0 every second – Line 36.
   (c) Operating Input X7 will Reset M0.
5. Line 15
   (a) The Data Registers D0–D3 are used as a look-up table which stores the timer values necessary for the following absolute drum instruction (ABSD), which is used to produce the waveforms in Figure 31.2.
   (b) The timer values stored in D0–D3 and the required action when that time is reached is shown in the table below:

Data register	Time (sec)	Action
D0	1	A+
D1	3	A−
D2	2	B+
D3	9	B−

Interrupts 321

(c) The registers in the look-up table operate in pairs. For example:
  (i) Register D0 stores the time delay value when Y0 is turned ON.
  (ii) Register D1 stores the time delay value when Y0 is turned OFF.
  (iii) Register D2 stores the time delay value when Y1 is turned ON.
  (iv) Register D3 stores the time delay value when Y1 is turned OFF.

6. Line 36
   (a) The Instruction -(C0 ^{K10})- is part of the ABSD instruction.
   (b) Its function is to count input pulses, in this case 1-sec pulses from M8013.
   (c) When the value in the counter reaches a value in the look-up table, then an output will be turned ON/OFF.
   (d) For example, after a delay of 1 sec, the value in C0 will be 1. This is also the value in D0 and hence Output Y0 turns ON and Piston A moves to its A+ position.

7. Line 41
   (a) The ABSD instruction -[ABSD D0 C0 Y0 K12]- is used to turn the Outputs Y0 and Y1 ON and OFF at the required intervals shown in the table on page 320.
   (b) The parts of this instruction are:
     (i) ABSD   Absolute Drum. This relates to a now-obsolete method for obtaining timed controller operations using a rotating mechanical drum.
     (ii) D0   This is the Head Address for the look-up table, which is storing the times required for outputs to turn ON/OFF.
     (iii) C0   Counter C0 is incremented every second from the 1-sec pulses of M8013. Therefore, at known times, the value in C0 will equal a value stored in the look-up table.
                When a value in C0 equals a value in the look-up table, then an output will be turned ON/OFF.
     (iv) Y0   The Output Y0 is the first output, which is being used in this sequence controller operation.
     (v) K2   The K2 value defines that the following is required:
                Two Outputs Y0 and Y1.
                A look-up table having 4 values.

8. Line 51
   (a) It can be seen from the output waveforms for Y0 and Y1 that the cycle time for turning the two pistons ON/OFF is 9 sec.
   (b) Hence, after 9 sec from the start of the cycle, the Instruction -[>= C0 K9]- will be TRUE.
   (c) Therefore, at this point the following take place:
     (i) The Cycle Counter D200 is incremented.
     (ii) Counter C0 is reset, which enables the start of the next cycle.
   (d) Note:
       D200, which is battery-backed, is used as the cycle counter to ensure that the cycle count is not reset, should there be a mains failure or the PLC is switched to STOP.

9. Lines 61, 65 and 69
   (a) When the interrupt occurs due to the closure of X0, there is the possibility that because of switch bounce, more than one interrupt could be generated (Ref. p254).

(b) Should a second ISR be executed due to switch bounce within a few milli-seconds of the first, then the following would happen if further X0 interrupts were not temporarily disabled:
   (i) The first ISR would display the contents of the cycle counter on the 7-segment displays and then reset D200.
   (ii) The second ISR would now display the reset contents of D200, i.e. 00.
   (iii) Since the cycle counter only displayed the correct count for a few milli-seconds before displaying 00, it would appear that no components had been produced.
(c) To overcome switch bounce, the following is carried out:
   (i) Line 69   At the start of the ISR, the Instruction -[SET M8050]- is executed, which disables further X0 interrupts from occurring.
   (ii) Line 61 After the ISR has been executed and the program returns to the main program, then on Line 61, the closure of the M8005 contacts starts T200 a 50 milli-second timer.
   (iii) Line 65 At the end of the time delay, the closure of the T200 contacts re-enables the X0 interrupt by re-setting M8050.
(d) Hence, it is not possible for more than one ISR to be executed within 50 milli-seconds of one other, even though the X0 Input may operate more than once, during this time.

31.7 Interrupt service routine

1. Line 69
   (a) Although the Interrupt Pointer I001 is entered on the ladder diagram, it will only display I1.

   The I001 is obtained as follows:
   I   0   0   1
   Interrupt X0   No meaning   Interrupt occurs on a positive going signal,
                                 i.e. X0 turning ON

   (b) To overcome the effects of switch bounce, the ISR initially executes the Instruction -[SET M8050]-, which disables the X0 interrupt.
   (c) The contents of the Cycle Counter D2000 are transferred in BCD format to the Outputs Y10–Y17 and displayed on the 7-segment displays.
   (d) The contents of D200 are then cleared.
   (e) The execution of the interrupt return instruction -[IRET]- indicates that the ISR has been completed and the program can return to the main program.

31.8 Monitoring – INT1

1. Initially, monitor the ladder diagram.
2. Let the program run automatically and when D200 displays a value of 8, operate the Interrupt button X0.

3. Use entry data monitor, to monitor the following:
   (a) D2000   The cycle counter.
   (b) K4Y10   The BCD output value.

Device	ON/OFF/Current	Setting value	Connect	Coil	Device comment
D200	0				End Of Cycle Counter
K4Y010	8				

Figure 31.3

4. As can be seen from Figure 31.3, with the operation of the X0 Input, an interrupt is generated, which carries out the following:
   (a) Transfers in BCD format the contents of the Cycle Counter D200 to the 7-segment displays connected to Y10–Y27.
   (b) Resets D200 back to 0.

32

Step counter programming

Step counter programming is a method of programming very large ladder diagrams, where it is essential that each part of the program operates in the correct sequence.

It is widely used in manufacturing industry to control sequential processes, i.e. the control of machine tools, assembly processes, etc.

It also enables very quick fault diagnosis by monitoring the step counter value, when the process suddenly stops operating.

The basic principle is as follows:

1. When the correct sequence of inputs occur at a point in the cycle, a step counter is incremented by 1.
2. The step counter is then used to turn ON a corresponding battery-backed M coil, known as a sequence control coil (see the following table).

Step counter value	Sequence control M coil
0	M500
1	M501
2	M502
3	M503
4	M504
5	M505
6	M506
7	M507
8	M508
9	M509
10	M510

3. The contacts of the sequence control M coils will then turn ON/OFF the required output or outputs for that part of the cycle.
4. The main disadvantage with this type of programming is that if at a later stage, modifications to the program are required it can be difficult to insert additional input contacts, unless a method known as counter assist is used. This method will be used in the project STEP_CNTR1.

32.1 Ladder diagram – STEP_CNTR1

```
 0    M8002                                              [RST   C0      ]
      ┤ ├                                                      STEP
      RUN                                                      COUNTER
      PULSE

      X1
      ┤/├
      STOP

      M510
      ┤ ├
      END OF
      CYCLE

 5    X1                                     *< Reset Y0-Y3
      ┤/├                                              [MOV   K0   K1Y0 ]
      STOP

*Step Coils
*M500-M510

11    M500   X0     X1     X2     X4        *<Counter Assist 1
      ┤ ├    ┤ ├    ┤ ├    ┤ ├    ┤ ├                              (M0   )
             START  STOP   A-     B-

      M501   X3     X4
      ┤ ├    ┤ ├    ┤ ├
             A+     B-

      M502   X3     X5
      ┤ ├    ┤ ├    ┤ ├
             A+     B+

      M503
      ┤ ├

      M504
      ┤ ├

27    M0                   M1               *<Counter Assist 2
      ┤ ├                  ┤/├                                     (M1   )

      M505   X2     X5
      ┤ ├    ┤ ├    ┤ ├
             A-     B+

      M506   T0
      ┤ ├    ┤ ├

      M507   X2     X4
      ┤ ├    ┤ ├    ┤ ├
             A-     B-

      M508
      ┤ ├

      M509
      ┤ ├
```

(Continued)

```
      M1                                              K10
43   ─┤ ├──────────────────────────────────────────( C0    )─
                                                    STEP
                                                    COUNTER

      M8000                                *<Convert C0 to M500-M515
47   ─┤ ├─────────────────────────────[ DECO   C0    M500    K4  ]─

      M500  X2    X4
55   ─┤ ├──┤/├──┤/├─────────────────────────────────( Y7    )─
       A-    B-                                     HOME
                                                    POSITION

      M501                                  *<A+
59   ─┤ ├───────────────────────────────────────[ SET    Y0    ]─

      M502                                  *<B+
61   ─┤ ├───────────────────────────────────────[ SET    Y1    ]─

      M505                                  *<A-
63   ─┤ ├───────────────────────────────────────[ RST    Y0    ]─

      M506                                             K50
65   ─┤ ├──────────────────────────────────────────( T0    )─
                                                    5 SEC
                                                    TIME
                                                    DELAY

      M507                                  *<B-
69   ─┤ ├───────────────────────────────────────[ RST    Y1    ]─

      M8000                              *<Display Step No.
71   ─┤ ├──────────────────────────────[ BCD    C0    K4Y10 ]─


77   ──────────────────────────────────────────────[ END   ]─
```

32.2 Principle of operation – STEP_CNTR1

1. The project STEP_CNTR1 is used to control two pistons A and B similar to PNEU4 (refer page 177).
2. Line 0
   (a) On switching the PLC to RUN, the contacts M8002 momentarily operate to enable the step counter C0 to be reset to zero.
   (b) With the stop switch OFF, its X1 normally closed contacts will be used to ensure the step counter C0 remains reset.
   (c) The M510 contacts are used to reset the step counter back to zero at the end of each cycle.

3. Line 5
   (a) With the stop button X1 not made, the X1 contacts will be closed and this will enable the instruction -[MOV K0 K1Y0]- to be executed.
   (b) This ensures that the outputs Y0 and Y1 remain OFF.
4. Line 47
   (a) The instruction -[DECO C0 M500 K4]- is essential to step counter programming.
   (b) The step counter C0 is decoded to turn ON a corresponding battery-backed sequence control M coil starting at M500.
   (c) The K4 value determines that only the first four bits of the C0 count value are to be decoded, i.e. when the value of C0 is in the range 0–15. However in this project, C0 will be reset back to zero when C0 reaches a value of 10.

Step counter value	Sequence control M coil
0	M500
1	M501
2	M502
3	M503
4	M504
5	M505
6	M506
7	M507
8	M508
9	M509
10	M510

5. Line 55
   (a) When C0 is reset to zero, then M500 will turn ON.
   (b) With the pistons not operated, then X2 and X4 will be closed and hence output Y7 will turn ON to indicate the system is at the home position.
6. Line 11 – M500
   (a) When the Start button X0 is operated, and with M500, X1, X2 and X4 already closed then M0, counter assist 1 will turn ON.
   (b) The program block, which connects to M0, consists of five lines of program connected in parallel.
   When all of the contacts in any one line are made, then M0 will be turned ON.
   Since each line in both blocks starts with a different M coil in the range M500 to M509, then only one line at a time can have all of its contacts closed.
7. Line 27
   Each time the M0 contacts close, then M1 counter assist 2, will also turn ON.
8. Line 43
   (a) With the M1 contacts closing, then C0 will be incremented by 1.
   (b) This will cause M500 to turn OFF and M501 to turn ON.
9. Line 59
   With the M501 contacts closing, output Y0 will be set ON and hence cause piston A to be energised.

10. Line 11 – M501
    (a) M501 now turns ON and when piston A is operated and travels to its outer limit, then limit switch X3 (A+) will close.
    (b) With X3 (A+) closing and with X4 (B−) already closed, M0 will again turn ON.
11. Line 27 – M0.
    (a) With the M0 contacts closing, this will again energise M1.
    (b) The M1 normally closed contacts ensure that the supply to C0 will break for a least one scan time.
    (c) This ensures that when a line is selected, i.e. M503 and that line is already complete, then the counter will not remain latched ON.
12. Line 43
    (a) With the M1 contacts closing, counter C0 will now be incremented to 2.
    (b) This in turn will cause M501 to turn OFF and M502 to turn ON.
13. Line 61
    With the closure of the M502 contacts output Y1 will now be set and piston B will travel to its B+ position.
14. Line 11 – M501 and M502
    (a) The procedure now continues with the first block of contacts.
    (b) With M502 and X3 already closed, then when piston B reaches its outer limit and X5 closes, M0 will once more turn ON.
    (c) This will cause C0 to be incremented to 3.
15. Line 11 – M503 and M504
    (a) The use of M503 and M504 enables modifications to be inserted at these points in the sequence, without affecting the existing sequence of operations.
    (b) With the closure of M503, then immediately M0 and M1 will turn ON. C0 will be incremented to 4 and hence M503 will open and M504 will close.
    (c) With the closure of M504, then immediately C0 will be incremented to 5 and hence M504 will open and M505 will close.
16. Line 63
    With the closure of M505, output Y0 will be reset and piston A will return to its A− position.
17. Line 27
    (a) With C0 having reached a value of 5, then M0 cannot be energised during the remainder of the cycle. Hence, only the second block of contacts, i.e. M505–M509, will now be used.
    (b) Therefore since M0 is no longer used in the counter assist process then for the remainder of the cycle only counter assist 2 will be used to increment C0.
    (c) At this moment in the cycle C0 = 5 and the M505 contacts are closed.
18. Line 27 – M505
    (a) When piston A returns to its A− position, then X2 closes.
    (b) As the M505 and the X5 contacts are already closed, M1 will now be energised and increment C0 to 6 – (Line 43).
    (c) Hence the M505 contacts will open and the M506 contacts will close.
19. Line 65
    (a) With C0 now having a value of 6, then the M506 contacts will close.
    (b) The 5-sec timer T0 will now be energised and start timing out.

20. Line 27 – M506
    (a) After 5 sec have elapsed, timer T0 will time out and its contacts will close.
    (b) The closure of the T0 contacts and with the M506 contacts already closed will enable M1 to be energised and hence cause step counter C0 to be incremented to 7.
    (c) Therefore the M506 contacts will open and the M507 contacts will close.
21. Line 69
    (a) With C0 now having a value of 7, then the M507 contacts will close.
    (b) The closure of the M507 contacts will cause output Y1 to be reset and hence piston B will return to its B− position.
22. Line 27 – M507
    When piston B reaches its B− position, then its X4 contacts will close and with the X2 and the M507 contacts already closed will enable M1 to be energised and hence cause step counter C0 to be incremented to 8.
23. Line 27 – M508 and M509
    (a) Since there are no input contacts associated with M508 then M1 will be immediately energised and cause C0 to be incremented to 9.
    (b) Hence contacts M508 will open and M509 will close.
    (c) Similarly since there are no input contacts associated with M509 then M1 will be immediately energised and cause C0 to be incremented to 10.
    (d) Hence contacts M509 will open and M510 will close.
    (e) The closure of the M510 contacts indicates that one full cycle has been completed and the next cycle can commence.
24. Line 0 – M510
    (a) With the M510 contacts closing, then step counter C0 will be reset back to zero.
    (b) Hence contacts M510 will open and M500 will close.
    (c) With the M500 contacts closing, the next cycle of operations can commence.
25. Line 71
    (a) The instruction -[BCD C0 K4Y10]- is used to display the contents of step counter C0.
    (b) Should a fault occur, i.e. a limit switch does not operate at the correct point in the machine cycle, this will cause the process to stop.
    (c) From the step counter display, the following can be determined immediately.
        i. Which M coil is still ON.
           For example, if C0 = 5, then M505 must be ON.
        ii. The line number for the M coil still ON can be found by using Find device in the find/replace menu.
        iii. From monitoring that particular line, determine which input has not operated.

32.3 Simulation and monitoring procedure

1. **Display and monitor the Ladder Diagram STEP_CNTR1.**
2. **Operate the Input Switches X1, X2 and X4.**
   C0 = 0
   This will simulate the operation of the stop push button and the A− and B− limit switches.

330 *Mitsubishi FX Programmable Logic Controllers*

   Since C0 is at 0, M500 will be made and hence Y7, the Home position LED will be ON.
3. **Momentarily operate the Start Switch X0.**
   C0 = 1
   Output Y0 will now be energised and this will cause piston A to operate to the A+ position.
   With piston A moving forward, limit switch X2 will open and limit switch X3 will close.
4. **Open X2 and Close X3.**
   C0 = 2
   This will cause Output Y1 to energise and hence enable piston B to move forward to its B+ position.
5. **Open X4 and Close X5.**
   C0 = 5
   M505 will now be ON and this will reset output Y0.
   Hence piston A will return to its A− position.
6. **Open X3 and Close X2.**
   C0 = 6
   With X2 closing, this will start the operation of the Timer T0.
7. **After 5 sec, Timer T0 will time out.**
   C0 = 7
   The closure of M507 will reset output Y1.
   Hence piston B will return to its B− position.
8. **OPEN X5 and Close X4.**
   C0 = 0
   The process can now be repeated, by momentarily operating X0.

32.4 Entry data monitoring – STEP_CNTR1

1. Monitor the program STEP_CNTR1 using entry data monitor, as shown in Figure 32.1.

Device	ON/OFF/Current	Setting value	Connect	Coil	Device comment
C0	0	10	0	0	Step Counter
M500			1		
M501			0		
M502			0		
M503			0		
M504			0		
M505			0		
M506			0		
M507			0		
M508			0		
M509			0		
M510			0		End of Cycle

Figure 32.1

2. Note:
   (a) After each step, the step counter C0 will be incremented by 1.
   (b) The sequence M coils will open and close one after the other.
   (c) At the end of the cycle when piston B returns to its B− position and X4 is remade, then C0 will return to zero and M500 will close.

32.5 Pneumatic panel operation

The PLC can now be connected once more to the SMC pneumatic panel (refer page 98) to enable the complete system to be tested.

This enables the PLC and the program STEP_CNTR1 to control more of an industrial type process than just being simulated with switches and LEDs.

Pneumatic drawing − STEP_CNTR1

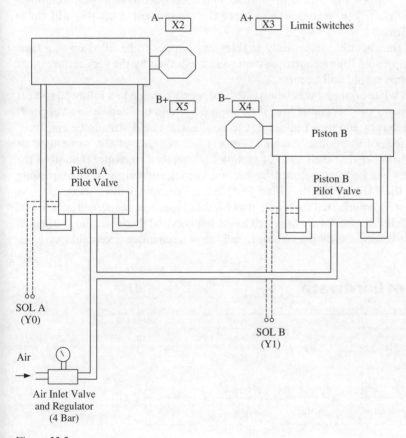

Figure 32.2

33

Automatic queuing system

This application is based on polycarbonate components being manufactured by five separate injection moulding machines.

The raw material is polycarbonate granules, which are stored in a central hopper and which are transferred to each machines' individual hopper by a vacuum system.

Whenever one of the hopper contents falls to a low value, then a signal is sent to the control system to enable the required hopper to be topped up with granules.

However if the hopper is not filled up in time, and consequently its injection moulding machine has insufficient granules to produce the component, then this will cause production problems.

The design of the vacuum system only enables one hopper to be filled up at a time and if a simple sequential filling control system were used, then by the very nature of its operation, problems would still occur.

If for example, while hopper 2 was being filled, hopper 1 became low followed a short time later by hopper 5 also becoming low, then hopper 5 would be filled before hopper 1. This could mean that by the time hopper 1 is to be filled, it could already be empty.

Should a number of moulding machine hoppers signal either at the same time or shortly after one another that their level of granules is low, then to ensure that all of the individual hoppers will be filled before they become completely empty, a PLC queuing system based on the FIFO (First In – First Out) Stack Principle, will be used.

The queuing system ensures that the order, in which the requests are made, will be identical to the order in which the granules are delivered to the hoppers and therefore, the granules in any of the individual hoppers will not be able to fall below a minimum acceptable level.

33.1 System hardware

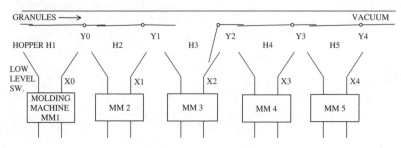

Figure 33.1

33.2 FIFO memory stack

The PLC software is based on a 5-input queuing system in which five inputs X0–X5 are scanned in turn approximately every 0.1 sec and where an input is ON, its corresponding logical value is stored in a FIFO memory stack.

As each logical value is removed from the stack, it will cause the corresponding Y output to turn on for 30 sec.

Input	Logical value	Y-Output
X0	1	Y0
X1	2	Y1
X2	4	Y2
X3	8	Y3
X4	16	Y4

33.3 Software diagram

The block diagram shown in Figure 33.2 shows the major software modules required for the complete system.

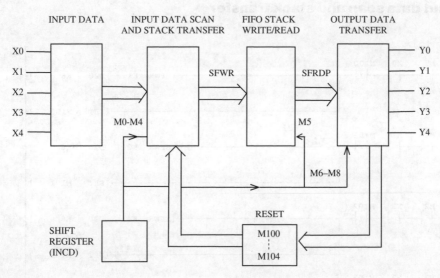

Figure 33.2

Basic operation

1. When an input is turned ON, its equivalent logical value is stored in the next available position on the FIFO stack.
2. After any one of the Y outputs has been ON for 30 sec, that particular output will turn OFF and the logical value which is at the top of the stack is now transferred to the output section.

3. The logical value is now used to turn ON the required Y output.
4. At the same time, the contents of the stack are all moved up one position.
5. The operation continues until all of the logical values stored on the stack have been shifted out. This will occur if none of the X inputs are ON and hence no new values are being written to the stack.

33.4 Ladder diagram – QUEUE1

Shift register

```
      M8000
0     ─┤ ├──────────────────────────────[FMOV  K1    D0    K9  ]─
                                       ─[INCD  D0    C0    M0    K9  ]─
      M8011                                                 K9999
      ─┤ ├──────────────────────────────────────────────────(C0    )─
```

Input data scan and stack transfer

```
       M0   X0   M100
21    ─┤ ├──┤ ├──┤/├────────────────────[SFWR  K1    D20   K6  ]─
                                                ─[SET   M100 ]─

       M1   X1   M101
32    ─┤ ├──┤ ├──┤/├────────────────────[SFWR  K2    D20   K6  ]─
                                                ─[SET   M101 ]─

       M2   X2   M102
43    ─┤ ├──┤ ├──┤/├────────────────────[SFWR  K4    D20   K6  ]─
                                                ─[SET   M102 ]─

       M3   X3   M103
54    ─┤ ├──┤ ├──┤/├────────────────────[SFWR  K8    D20   K6  ]─
                                                ─[SET   M103 ]─

       M4   X4   M104
65    ─┤ ├──┤ ├──┤/├────────────────────[SFWR  K16   D20   K6  ]─
                                                ─[SET   M104 ]─
```

Output data transfer

```
      M5    M50
76   ─┤├───┤/├──────────────────────────[SFRDP  D20   D30   K6  ]─

                                        ─[MOV   D30   K2Y0      ]─

      Y0    M50   M6
90   ─┤├───┤/├──┤├────────────────────────────────[SET    M50  ]─
      Y1
     ─┤├──
      Y2
     ─┤├──
      Y3
     ─┤├──
      Y4
     ─┤├──

      M50                                                  K300
98   ─┤├──────────────────────────────────────────────(T0       )─
            T0    M7
           ─┤├───┤├────────────────────────[MOV    K0    D30   ]─

109  ─[=  D30   K0 ]─────────────────────────────────(M10      )─

      M10   M8
115  ─┤├───┤├─────────────────────────────────────[RST    M50  ]─
```

Reset

```
        Y0
118    ─┤↓├──────────────────────────────────────[PLF  M110  ]─

        M110
121    ─┤ ├──────────────────────────────────────[RST  M100  ]─

        Y1
123    ─┤↓├──────────────────────────────────────[PLF  M111  ]─

        M111
126    ─┤ ├──────────────────────────────────────[RST  M101  ]─

        Y2
128    ─┤↓├──────────────────────────────────────[PLF  M112  ]─

        M112
131    ─┤ ├──────────────────────────────────────[RST  M102  ]─

        Y3
133    ─┤↓├──────────────────────────────────────[PLF  M113  ]─

        M113
136    ─┤ ├──────────────────────────────────────[RST  M103  ]─

        Y4
138    ─┤↓├──────────────────────────────────────[PLF  M114  ]─

        M114
141    ─┤ ├──────────────────────────────────────[RST  M104  ]─

143    ─────────────────────────────────────────────[END    ]─
```

33.5 Principle of operation – QUEUE1

Shift register

Line 0

1. The shift register consists of nine memory coils M0–M8, which are sequentially turned ON for 0.01 sec and then OFF for 0.08 sec. The instruction, which is used for this purpose is the incremental drum INCD.
2. Initially the data registers D0–D8 are all filled with the value of 1, using the instruction -[FMOV K1 D0 K9]-.
3. Each time the value in counter C0 reaches the value stored in each data register D0–D8, i.e. a 1, then the corresponding memory coil M0–M8, will turn ON. For example, when C0 reaches the value in D0, i.e. a 1, then M0 turns ON.
4. Counter C0 is automatically reset and then starts recounting. When the value in C0 reaches the value in D1, i.e. as before a 1, then M0 turns OFF and M1 turns ON.

5. The process constantly repeats from M0 to M8 and then back to M0.
6. The 0.01-sec pulses from M8011 are used to increment C0.

Shift register waveforms

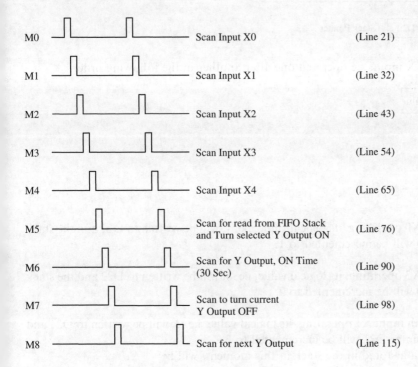

M0	Scan Input X0	(Line 21)
M1	Scan Input X1	(Line 32)
M2	Scan Input X2	(Line 43)
M3	Scan Input X3	(Line 54)
M4	Scan Input X4	(Line 65)
M5	Scan for read from FIFO Stack and Turn selected Y Output ON	(Line 76)
M6	Scan for Y Output, ON Time (30 Sec)	(Line 90)
M7	Scan to turn current Y Output OFF	(Line 98)
M8	Scan for next Y Output	(Line 115)

Input data scan and stack transfer

Lines 21–65

1. When any one of the inputs X0–X4 is turned on, then on being scanned by its corresponding shift register output, the logical value K_L for that input is written to the FIFO stack. This is done by, using the using the shift write instruction -[SFWR K_L D20 K6]-.
2. This instruction determines the following:
   (a) The Stack will consist of six data registers D20 to D25.
   (b) The stack pointer will be D20.
   (c) The stack registers which are to be used for storing the logical values, will be D21–D25.
   (d) When writing to the stack, it is the stack pointer's contents, which are used to calculate, which stack register will be used.
   (e) For example if D20 = 3, then stack register D(21 + 3), i.e. D24, will be where the next logical value will be written to.
   (f) Each time a logical value is written to the stack, the stack pointer is incremented by one.

3. Stack write format

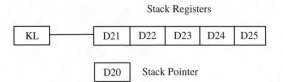

4. Let the five X inputs be operated one after another in the following order.

Input	Logical value
X0	1
X3	8
X1	2
X4	16
X2	4

5. Line 21
   With input X0 operating, its logical value, i.e. 1, will be written to D21 and the stack pointer D20 will be incremented to 1.
6. Line 32
   With input X3 operating, its logical value, i.e. 8, will be written to D22 and the stack pointer D20 will be incremented to 2.
7. Line 43
   Similarly with input X1 operating, its logical value, i.e. 2, will be written to D23 and the stack pointer D21 will be incremented to 3.
8. Hence the data stored in the stack at this moment, will be:

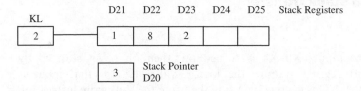

9. Note: The next stack register available for writing to, is D24.
10. The process continues until all of the five logical values have been stored onto the stack.

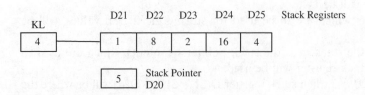

11. Line 21
    After the logical value for input X0 has been written to the stack, memory coil M100 is set and its associated normally closed contacts will open.
    This ensures that while the logical value corresponding to X0 is stored on the stack, the same value cannot be rewritten to the stack, even if X0 stays operated.
    This facility therefore inhibits a logical Value being written to the Stack, while that value remains stored on the Stack.

Line no.	Stack inhibit memory
21	M100
32	M101
43	M102
54	M103
65	M104

FIFO stack – read

Line 76

1. Data is read from the stack to data register D30, each time the shift read instruction -[SFRDP D20 D30 K6]- is executed.
2. Stack read format

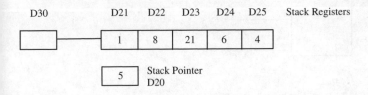

3. Each time the shift register contact M5 closes, the contents at the top of the stack, i.e. the contents of D21 are transferred to data register D30.
4. At the same time, the contents of the remaining stack registers are then shifted left one place and the contents of the stack pointer D20 are decremented by one.
5. The first time the stack read instruction -[SFRDP D20 D30 K6]- is executed, the contents of the stack will become as shown below.

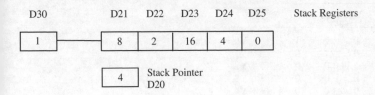

6. The second time the stack read instruction is executed, the contents of the stack are again shifted left one place and will become, as shown below.

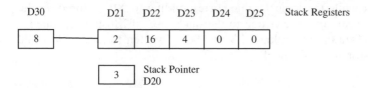

Output data transfer

Line 76
After the execution of each shift register read instruction, which transfers the logical value stored at the top of the stack to D30, the output transfer instruction -[MOV D30 K2Y0]- is now executed.

This enables the contents of data register D30 to be transferred to the actual Y outputs (see Figure 33.3).

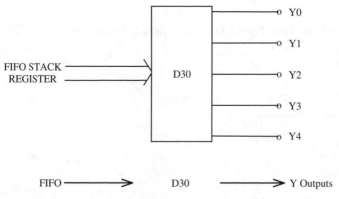

Figure 33.3

Output truth table

The Y outputs are turned ON, according to the following truth table.

Logical value	Binary					Y Output ON
	Y4	Y3	Y2	Y1	Y0	
01	0	0	0	0	1	Y0
02	0	0	0	1	0	Y1
04	0	0	1	0	0	Y2
08	0	1	0	0	0	Y3
16	1	0	0	0	0	Y4

Hence if the contents of D30 = 08, then output Y3 will turn ON.

Output – ON time

Lines 90–98
When a Y output is turned ON, it is only ON for 30 sec, this being sufficient time to fill an individual hopper with the polycarbonate granules.
   The following describes how this is done.
Line 90
With the operation of any one of the five Y outputs, M50 will be set, when the shift register output M6 turns ON.
Line 98
1. The closing of the M50 contacts will start a 30-sec time delay T0.
2. At the end of the 30 sec, the T0 contacts will close and with the operation of the shift register output M7, the instruction -[MOV K0 D30]- will be executed.
3. This instruction clears the contents of D0 and on the next scan of the program, when the instruction -[MOV D30 K2Y0]- is executed (Line 76), the Y output which is currently ON will be turned OFF.

Next Y output

Line 109
1. When a particular Y output has completed its hopper fill sequence, then it is necessary for the next Y output to be turned ON.
2. The logical value for the next Y output is currently at the top of the stack and hence the stack read instruction has now to be executed.
3. With D30 being cleared at the end of the 30-sec time period, a comparison is now made of the contents of D30, using the instruction -[=D30 K0]-.
Line 115
With D30 = 0, then M10 will operate and this plus the shift register output M8, will reset M50.
Line 76
1. With the closure of the M50 normally closed contacts, the stack read instruction -[SFRD D20 D30 K6]- and the output transfer instruction -[MOV D30 K2Y0]- will both be executed.
2. This will enable the next logical value stored at the top of the stack to be transferred to D30 and hence the next required Y output, will be turned ON.

Reset

Lines 118–141
Line 118
1. After an output, i.e. Y0 has been turned ON and then OFF, the normally open contact of Y0 will reopen to enable the instruction PLF M110 to be executed.
2. PLF waveforms (see Figure 33.4).
3. On the falling edge of the Y0 contact, the instruction PLF 110, will be executed. This instruction causes the operation of M110 for one scan time, i.e. it is similar to PLS, but whereas PLS operates on the rising edge of an Input (refer page 83) PLF operates on the falling edge of an Input.

PLF waveforms

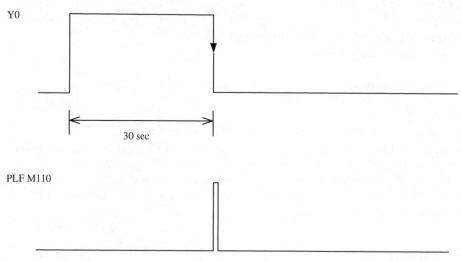

Figure 33.4

4. This in turn causes M110 to output a pulse for 1 scan time, which resets M100, the Stack inhibit memory and re-enables the X0 input in the input data scan and transfer section (Line 21).
5. Should X0 still be ON or come ON, then the transfer of the logical value corresponding to X0, will be reloaded back onto the Stack.
6. The other stack inhibit memories, i.e. M101, M102, M103 and M104 will also be reset, when their corresponding Y output is turned OFF.

33.6 Testing – QUEUE1

To test that the queue1 system operates as specified, carry out the following:

1. Ensure all of the X inputs and the Y outputs are OFF.
2. Switch On the X inputs in the following order and leave them switched ON. X0, X3, X1, X4, X2.
3. The Y outputs will now come on one at a time, every 30 sec, in the following order. Y0, Y3, Y1, Y4, Y2.
4. The cycle will repeat itself until the X inputs are switched OFF, or the RUN is switched OFF.

33.7 Monitoring – QUEUE1

1. Using entry data monitor, register and then monitor the devices listed in Figure 33.5.
2. The monitor details are for when output Y3 has just turned ON.

Device	ON/OFF/Current	Setting value	Connect	Coil	Device comment
T0	7	300	0	1	
D20	4				
D21	2				
D22	16				
D23	4				
D24	1				
D25	0				
D30	8				
Y000			0		
Y001			0		
Y002			0		
Y003			1		
Y004			0		

Figure 33.5

33.8 Analysis of results

From the monitor details, the following can be determined:

1. The value in D30 = 8 and hence the binary pattern which has been sent to the Y outputs is 0000 1000, i.e. bit 3 = 1.
2. This corresponds to Y3 being ON.
3. This is correct since the output Y3 was ON, when the monitored results were taken.
4. From the monitored results, the contents of the stack will be as shown in the diagram below.

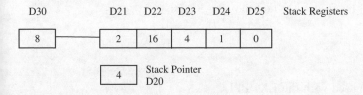

This shows the following:

1. The contents of D30 = 8, i.e. the logical value for the output Y3.
2. The stack pointer = 4, indicating there are four values left on the Stack.
3. The values which remain on the stack and their corresponding Y outputs are:

    2   16   4   1
    Y1 Y4 Y2 Y0

34

Analogue to digital conversion FX2N-4AD

34.1 Introduction

In process control systems many of the input signals are from analogue type devices known as transducers. Transducers are used to provide a voltage or current signal proportional to the measured value of, for example:

(a) Mass
(b) Linear/angular movement
(c) Velocity
(d) Temperature
(e) Pressure
(f) Strain.

Therefore before a signal from a transducer can be used by a PLC, it has to be converted into a digital quantity, using an analogue to digital converter – ADC.

The type of ADC, which will be used in this section, is the Mitsubishi FX2N-4AD. The FX2N 4ADC can be used with the following PLCs:

(a) The FX2N
(b) The FX1N.

However, it cannot be used with the FX1S PLC.

34.2 FX2N-4AD buffer memory addresses and assignments

Within the FX2N-4AD, is a Buffer Memory – BFM, which consists of 32 buffer memory locations.

The following table lists the addresses of each BFM location and its associated assignment.

Analogue to digital conversion FX2N-4AD

BFM	Contents
*0	Channel initialisation Default = H0000
*1	Number of samples for averaging for CH1 (1–4096) Default = 8
*2	Number of samples for averaging for CH2 (1–4096) Default = 8
*3	Number of samples for averaging for CH3 (1–4096) Default = 8
*4	Number of samples for averaging for CH4 (1–4096) Default = 8
5	CH1 averaged value
6	CH2 averaged value
7	CH3 averaged value
8	CH4 averaged value
9	CH1 present value
10	CH2 present value
11	CH3 present value
12	CH4 present value
13–14	*Reserved for future use*
*15	Conversion Speed 0 = 15 ms per channel 1 = 6 ms per channel
16–19	*Reserved for future use*
*20–24	Offset/Gain adjustment using software
25–28	Reserved for future use
29	Error status
30	Identification Code (K2010)
31	*Cannot be used*

TO instruction

For the buffer memory locations that have a '*' mark, data can be written to those locations by using the 'TO' instruction.

This information must be written to allow the unit to perform the correct conversion.

FROM instruction

For the buffer memory locations without the '*' mark, data can be read from those locations to the CPU, using the 'FROM' instruction.

34.3 Voltage and current conversion

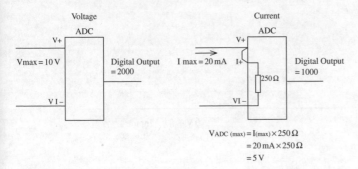

Figure 34.1

Voltage conversion

From Figure 34.1, it can be seen that an FX2N-4AD, when supplied with a maximum input voltage of +10 V, will convert that voltage to a maximum digital output of +2000.

Current conversion

When used for current conversion, the ADC unit has a 250 Ω resistor connected between the I+ and the VI− terminals and it is the potential difference across the 250 Ω resistor which is being converted.

Hence for a maximum input current of +20 mA, the actual voltage applied to the ADC will be only 5 V, which will produce a factory setting default output of only 1000.

Therefore, to obtain the maximum digital output of 2000, the gain value of the ADC will have to be reduced by a factor of 2 (refer page 348).
Note: To obtain current conversion it is necessary to place a link between the V+ and the I+ terminals. This is because the ADC electronics are connected to the V+ terminal, whilst the 250 Ω resistor is connected between the I+ and the VI− terminals.

34.4 Resolution − maximum input voltage

The voltage resolution of an ADC unit, is defined as the change in input voltage, which will produce a one-bit change in the digital output.

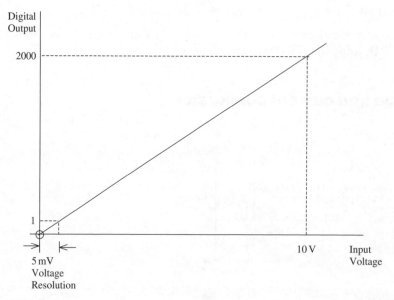

Figure 34.2

Maximum digital output FX2N-4AD

The maximum digital output from a Mitsubishi FX2N-4AD is −2047 to +2048.

Resolution 0–10V

Where a transducer has a maximum output of 10 V, i.e. the maximum input voltage into the ADC, then the resolution will be:

Voltage resolution (max. input voltage) = $\dfrac{\text{Maximum input voltage}}{\text{Maximum digital output}}$
The maximum input voltage = 10 V
The maximum digital output = 2000
Voltage resolution (max. input voltage) = $\dfrac{10}{2000}$ V = 5 mV

34.5 Resolution – maximum input current

The maximum input current = 20 mA
The maximum digital output = 2000
Current resolution (max. input current) = $\dfrac{20}{2000}$ mA = 10 µA

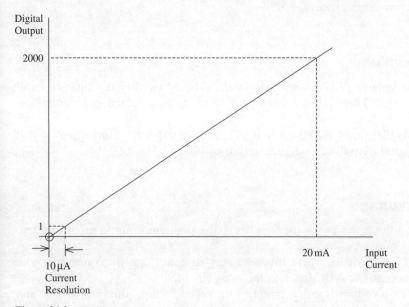

Figure 34.3

Note: The voltage or current resolution of a Mitsubishi ADC can be improved, by increasing the slope of the graph shown in Figure 34.3.
This is done by reducing the gain value of the ADC (refer page 348).

34.6 Relationship between Vin and digital output

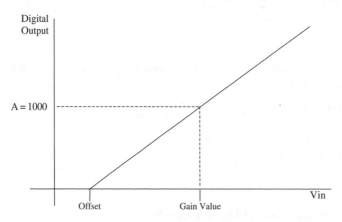

Figure 34.4

The graph in Figure 34.4 shows the relationship between the analogue input voltage and the digital output value. The digital output will be stored in a data register within the CPU, after the conversion process.

Voltage gain value

The user sets the voltage gain value to enable the ratio of the digital output to input voltage to be changed. This enables a full digital output to be obtained, even with a low input voltage.

It is defined as that input voltage, which will produce a digital output equal to half the maximum digital output, i.e. a digital output of 1000.

Input offset voltage

This is the input voltage, which the user applies to the analogue input to produce a zero digital output.

It is very useful in systems, where it is necessary to detect that the connecting wire from the transducer to the ADC has become open circuited.

Note: Since the gain voltage is independent of the offset voltage, the gain voltage and the offset voltage can be regarded as two separate points, on the straight-line ADC graph.

34.7 ADC equations

The equation for the straight-line graph as shown in Figure 34.5, is of the form $y = mx + c$.

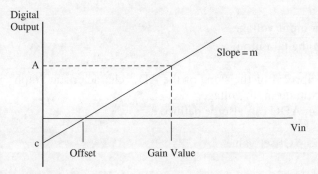

Figure 34.5

Slope – m

The slope of the ADC graph $(m) = \dfrac{\text{Change in digital output}}{\text{Change in input voltage}}$

$$m = \dfrac{A}{\text{Gain value} - \text{Offset}}$$

Constant – c

The constant c is obtained from

$$\text{Slope} = \dfrac{c}{\text{offset}}$$

From the graph it can be seen that if the Offset is a positive value, then c must be negative.

i.e. $c = -(\text{Slope} \times \text{Offset value})$

Complete equation

Hence using the equation for a straight-line graph, i.e. $y = mx + c$

$$\underset{(y)}{\text{Digital output}} = \underset{(m)}{\text{Slope}} \times \underset{(x)}{\text{Vin}} - \underset{(\,c\,)}{\text{Slope} \times \text{Offset}}$$

$$= \text{Slope} \times (\text{Vin} - \text{Offset})$$

$$= \dfrac{A}{\text{Gain value} - \text{Offset}} \times (\text{Vin} - \text{Offset})$$

$$\text{Digital output} = A \times \dfrac{\text{Vin} - \text{Offset}}{\text{Gain value} - \text{Offset}}$$

34.8 Resolution – independent of input voltage

The definition for the resolution of an ADC is the change in input voltage, which will produce a change of one digit in the digital output.

$$\text{Resolution} = \frac{\text{Change in input voltage}}{\text{Change in digital output}}$$

As can be seen this is the reciprocal of the slope of the ADC characteristic graph, which is independent of the maximum input voltage.

Therefore, the resolution of an ADC can also be defined as:

$$\text{Resolution} = \frac{\text{Gain value} - \text{Offset value}}{A}$$

Input voltage limitations

If the offset voltage and the voltage gain value were set to the same value, then when the input voltage reaches the offset/gain value, the digital output characteristic would be a straight vertical line until the it reaches its maximum value, and would stay there for any additional increase in input voltage (see Figure 34.6).

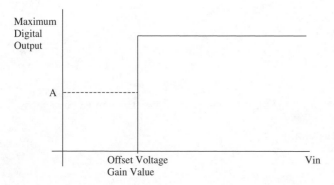

Figure 34.6

Under such circumstances, it would be impossible to determine what the digital output value would be, since a very small change in input voltage would produce a very large change in the output value. To overcome this problem, Mitsubishi specify the minimum difference between the offset voltage and gain value must be equal to or greater than 1 V.

$$(\text{Gain value} - \text{Offset voltage}) >= 1\,\text{V}$$

Input current limitations

Similarly the input offset current and the current gain value cannot be identical. This difference should be equal to or greater than 4 mA.

(Current gain value − Offset value) >= 4 mA

Resolution ranges

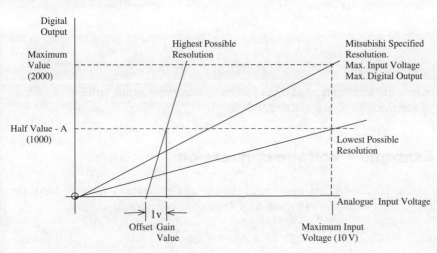

Figure 34.7

34.9 Highest possible resolution

If it is required to measure an input signal to the highest possible accuracy, i.e. set the ADC unit such that it can measure the smallest change in the input signal, then

Gain value − Offset value = 1 V

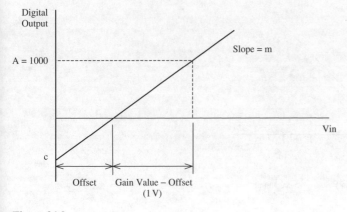

Figure 34.8

$$\text{Resolution} = \frac{1}{\text{Slope of above graph}}$$

$$= \frac{\text{Gain value} - \text{Offset}}{A}$$

Gain value − Offset = 1 V minimum

A = 1000

$$\text{Hence highest possible resolution} = \frac{1\,\text{V}}{1000}$$

$$= 1.0\,\text{mV}$$

Note: Refer page 347.
This compares with a resolution of 5 mV, when a maximum input voltage of 10 V is converted to a maximum digital output of 2000.

34.10 Example – voltage conversion

Channel 1 of the FX2N-4AD is to be used for voltage conversion and will have the following parameters when the program ADC1 is executed (refer page 362).

(a) A = 1000
(b) Offset voltage is +2 V
(c) Gain value is +6 V.
See Figure 34.9 below

Determine:

(a) The digital output when Vin = +8 V (answer on page 353).
(b) The digital output should the input signal become disconnected.
(c) The resolution.

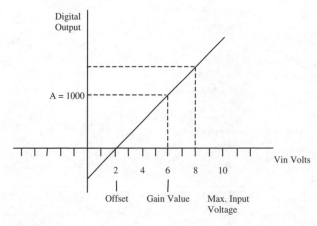

Figure 34.9

Digital output (Vin = 8 V)

$$\text{Digital output (Vin = 8 V)} = A \times \frac{\text{Vin} - \text{Offset}}{\text{Gain value} - \text{Offset}}$$

$$= 1000 \times \frac{8\,V - 2\,V}{6\,V - 2\,V}$$

$$= 1000 \times \frac{6\,V}{4\,V}$$

$$\text{Digital output} = 1500$$

34.11 Example – current conversion

Channel 2 of the FX2N-4AD is to be used for current conversion and will have the following parameters, when the ADC program is executed.

(a) A = 1000.
(b) Offset voltage is +4 mA.
(c) Maximum input current is 20 mA.
See Figure 34.10 below

Determine the following:

(a) The gain value.
(b) The digital output when Iin = +7 mA (answer on page 354).
(c) The digital output, should the input become disconnected.
(d) The actual resolution (answer on page 354).

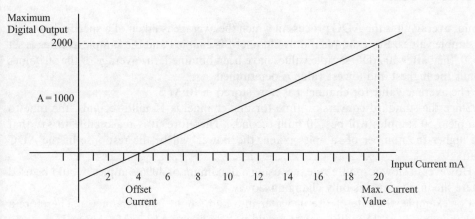

Figure 34.10

Digital output

$$\text{Digital output} \atop (\text{In} = 7\,\text{mA}) = A \times \frac{\text{Iin} - \text{Offset}}{\text{Gain value} - \text{Offset value}}$$

$$= 3000 \times \frac{7\,\text{mA} - 4\,\text{mA}}{12\,\text{mA} - 4\,\text{mA}}$$

$$= 3000 \times \frac{3\,\text{mA}}{8\,\text{mA}}$$

Digital output = 375

Current resolution

$$\text{Current resolution} = \frac{\text{Gain value} - \text{Offset value}}{A}$$

$$= \frac{12\,\text{mA} - 4\,\text{mA}}{1000}$$

$$= \frac{8\,\text{mA}}{1000}$$

Current resolution = $8.0\,\mu\text{A}$

This means that if the input current changes by $8.0\,\mu\text{A}$, then the digital output value will change by 1 digit.

34.12 Count averaging

Count averaging is the ADC process in which the average is taken of a specified number of sample values. For example if only channel 1 is enabled and the sample average is set to 50, then after 50 ADC sample values have been obtained, the average of the 50 values minus the highest and lowest value is determined.

The average value for channel 1 is now stored in BFM 5.

Since the standard conversion time for one channel is 15 milli-seconds, the time to complete 50 samples will be 750 milli-seconds. Therefore, it is reasonable to say that the higher the number of samples taken, the slower will be the response of the ADC unit.

However, a slow response rate, i.e. using a maximum of 4096 samples, would be used where the input signal is only changing slowly.

For example, the rate of rise of water in a dam could be hours/metre. Therefore a measurement every 15 milli-seconds would not show any noticeable change.

34.13 Positioning the analogue unit

Figure 34.11 shows how the analogue unit is incorporated into an FX2N System:

```
                    Block Zero
┌─────────────────────────┬──────────────┐
│  INPUT TERMINALS        │              │
│                         │              │
│      MITSUBISHI         │   POWER ▯    │
│                         │              │
│      MELSEC FX2N        │  FX2N – 4AD  │
│                         │              │
│  OUTPUT TERMINALS       │   24V ○      │
│                         │   A/D ○      │
└─────────────────────────┴──────────────┘
```

Figure 34.11

The unit that is located in the first position to the right of the base unit is deemed to be block zero.

If required an additional seven special units can be added, making a possible total of eight special units (blocks 0–7) on any one system.

If required, the special function units available within the FX2N range can be mixed in any one system. Care must be taken when doing this as the current consumption of certain special function units exceeds that which is available from the base unit and an additional extension unit would have to be fitted.

34.14 ADC wiring diagram

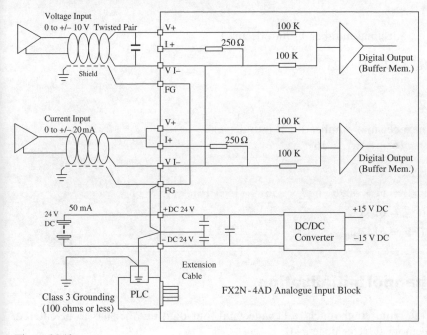

Figure 34.12

The unit should be connected as shown in Figure 34.12. However, there are five instances where care should be taken. These are listed below.

1. Twisted pair shielded cable should be used. This cable should also be isolated from power lines, which could induce noise.
2. If noise (electrically induced) is present at the input, then a smoothing capacitor can be used. The rating of which should be 0.1–0.47 µF.
3. You MUST when using a current input, connect terminals V+ to I+.
4. If there is excessive noise, connect the FG terminal with the electrically grounded terminal on the FX2N-4AD.
5. Connect the ground terminal on the FX2N-4AD unit to the ground terminal on the base unit.

34.15 Hexadecimal numbering system for special units

As is already known, the FX PLC is addressed in octal. However, when setting up the FX2N-4AD, it is necessary to use hexadecimal values. Hexadecimal or 'Hex' as it will be referred to from now on combines both numeric and alpha characters.

When counting in decimal or denary the numbering system is to the base 10. In octal, the numbering system is to the base 8. However in hexadecimal, the numbering system is to the base 16. An example of the first 16 hexadecimal numbers is shown below.

Decimal   0 1 2 3 4 5 6 7 8 9 10 11 12 13 14 15
Hexadecimal 0 1 2 3 4 5 6 7 8 9 A B C D E F

Example 1

Convert the hexadecimal number 17 into decimal.

$$(1 \times 16) + (7 \times 1) = 23$$

Example 2

Convert the hexadecimal number 17FF into decimal.
Power 16^3 16^2 16^1 16^0
Digits   1   7   F   F

$$\text{Total no. (Hex.)} = (1 \times 4096) + (7 \times 256) + (15 \times 16) + (15 \times 1)$$
$$= 6143$$

Hence $H17FF = 6143_{(10)}$

34.16 Channel initialisation

To initialise the required channels, a hexadecimal four-digit number has to be entered into buffer memory 0 (refer page 344).

The least significant character controls channel 1 whereas the most significant figure controls channel 4. If required, each channel can be set to accept a different input signal.

The example below (see Figure 34.13) demonstrates the following:

(a) Channels 1 and 2 are set for voltage and current respectively.
(b) Channels 3 and 4 are disabled.

Example

Data value 0: Pre-set range (−10 V to +10 V)
Data value 1: Pre-set range (+4 mA to +20 mA)
Data value 2: Pre-set range (−20 mA to +20 mA)
Data value 3: Channel off.

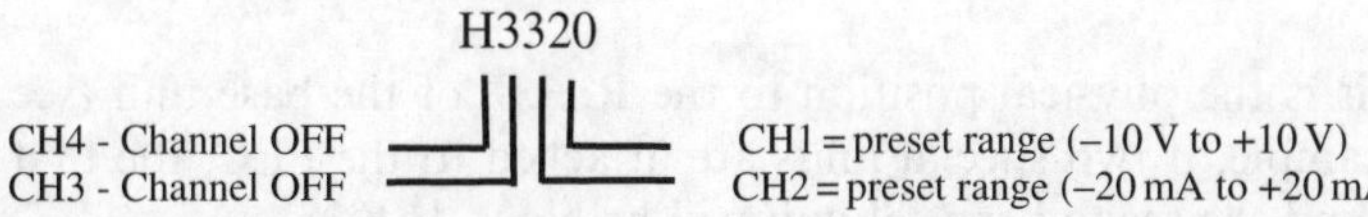

```
                        H3320
CH4 - Channel OFF  ┐ ┃ ┃ ┌   CH1 = preset range (−10 V to +10 V)
CH3 - Channel OFF  ┘ ┃ ┃ └   CH2 = preset range (−20 mA to +20 mA )
```

Figure 34.13

Hence if H3320 is entered into buffer memory 0, then:

(a) Channel 1 is set for −10 V to +10 V.
(b) Channel 2 is set for −20 mA to +20 mA.
(c) Channel 3 OFF.
(d) Channel 4 OFF.

34.17 TO and FROM instructions

To program the FX2N-4AD, it is necessary to use the 'TO' and 'FROM' instructions for transferring information between the CPU and the ADC.

1. The TO instructions are used to transfer the SET UP information to the ADC.
2. The FROM instructions are used to transfer:
   (a) The converted digital value from the ADC to a data register within the CPU.
   (b) The status of the ADC to the CPU, i.e. the notification of any errors within the ADC unit.

Block diagram

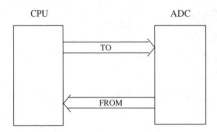

Figure 34.14

To

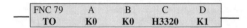

Figure 34.15

A   Block Location – This is the physical position to the R.H.S. of the base unit (see Figure 34.15). For example, if two special units are attached to the PLC, the first will be block 0 (K0), and the second special unit will be block 1 (K1).
B   Buffer memory area – This position writes to the required buffer memory location. For example, K0 means that channel initialisation data, i.e. H3320 will be transferred to buffer memory 0 (refer page 345).
C   Data requirements of the FX2N-4AD – Information in this location is sent to the buffer memory defined in B. In the above example, the data value H3320 is transferred to BFM 0.
D   Amount of information to be transferred – This indicates the amount of data to be transferred to the analogue unit. For example, as shown in the example above 'K1' indicates that only one word of information, i.e. H3220 is to be transferred to BFM 0.

From

FNC 78	A	B	C	D
FROM	K0	K5	D0	K4

Figure 34.16

A   Block Location – This is the physical position of the unit, to the R.H.S. of the base unit (see Figure 34.16). For example if two blocks are attached to the PLC, the first will be block 0 (K0), and the second block will therefore be block 1 (K1).
B   Buffer memory area – This position reads from the required buffer memory location. In the above example, the CH1-converted digital value, which is stored in BFM 5 will be transferred to D0.
C   Destination of data read by the analogue unit – In the example shown above, the converted digital output value for CH1 is transferred to data register D0.

D   Amount of information to be transferred – This indicates the amount of data to be transferred from the analogue unit. For example as shown in the example above, 'K4' indicates that four words of information are to be transferred to the PLC. Therefore with just one instruction the following would occur:

The contents of BFM 5 – CH1 – would be transferred to D0.
The contents of BFM 6 – CH2 – would be transferred to D1.
The contents of BFM 7 – CH3 – would be transferred to D2.
The contents of BFM 8 – CH4 – would be transferred to D3.

34.18 ADC errors – BFM 29

If an error or errors occur within the ADC, then by monitoring the contents of BFM 29 it is possible to determine what has occurred.

The instruction shown in Figure 34.17 is used to transfer the 16-bit contents of BFM 29 from the ADC to the M coils M10–M25, within the FX2N PLC.

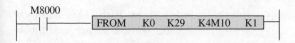

Figure 34.17

The resultant bit pattern when used with the table below will enable the error or errors to be determined.

Bit		ON	OFF	M Coil
b0:	Error	When any of b1 to b3 is ON, D/A conversion is stopped for the error channel	No error	M10
b1:	Offset/Gain error	Offset/gain data in EEPROM is corrupted or adjustment error	Offset/gain data normal	M11
b2:	Power abnormality	DC 24 V power supply failure	Power supply normal	M12
b3:	Hardware error	D/A converter or other hardware failure	Hardware normal	M13
b10:	Range error	Digital output value is outside the specified range (−2047 to +2048)	Digital output value is normal	M20
b11:	Averaging error	Number of Averaging samples is 4097 or more Number of Averaging samples 0 or less	Averaging is normal (between 1 and 4096)	M21
b12:	Offset/gain adjust prohibit	Prohibit–bits b1,b0 of BFM 21 are set to (1,0)	Permit – bits b1, b0 of BFM 21 are set to (0,1)	M22

Note:
Bits b4–b9, and bits b13, b14 and b15 are undefined.

34.19 Buffer memory – EEPROM

The values in the following buffer memories are also be copied to a non-volatile EEPROM memory within the FX2N-4AD.

  BFM 0 Channel initialisation.
  23 Offset value in mV or µA.
  24 Gain value in mV or µA.

The offset/gain values for each channel, which is written one channel at a time into BFM 23 and 24, are only transferred to the EEPROM, when the channel selection value is written to BFM 22.

EEPROM – Caution

When using the FX2N-4AD it is important that the following recommendations are taken into account.

1. The EEPROM has a life of approximately 10 000 cycles/changes, therefore it is recommended that programs which frequently change the contents of these EEPROM memory locations are not used.
2. Owing to the time required to actually write information to an EEPROM memory location, it is essential that a time delay of at least 300 mS is used before writing to the same or another EEPROM memory location.

34.20 Software programming of offset and gain

To enable the offset and gain values to be programmed into the ADC unit, the following buffer memories shown in the table below, are now used.

BFM	Contents	b7	b6	b5	b4	b3	b2	b1	b0
*20	Reset to defaults and presets	Default = 0							
*21	Offset, gain adjust	Default = (0,1) Permit changes							
*22	Offset, gain adjust	G4	04	G3	03	G2	02	G1	01
*23	Offset value	Default = 0							
*24	Gain value	Default = 5000							

The buffer memory numbers marked with a * indicate that they can be written from the PLC using the 'TO' command.

BFM 20: By writing a 1 to this area, all the settings are reset to default conditions. This can be used if incorrect data is entered from the PLC.
BFM 21: Default condition, i.e. (0,1), enables the writing of information to the buffer memory.
BFM 22: If for example, channel 1 is to be set up, then by writing a '1' to G1 and O1 the offset and gain values previously written into BFM 23 and 24, will now be transferred to the memory location for channel 1 within the EEPROM.

Similarly if channel 2 is to be set up, then by writing a '1' to G2 and O2, the new values set into BFM 23 and 24 will be now be transferred to the memory location for channel 2 within the EEPROM.

Note: The offset and gain values for each channel can only be set in steps of 5 mV or 20 μA.

34.21 Detecting an open circuit

If any channel is setup to convert an input current, and if it is essential that a disconnection be detected, then the range −20 mA to +20 mA range must be selected and not the 4 mA to 20 mA range. This is because the digital output characteristic of the 4–20 mA range does not go below zero, i.e. it does not output a negative value. Between 0 mA and 4 mA, the digital output value remains at 0. It will only rise above 0 when the input current exceeds 4 mA. Hence, it will not be possible to detect if the input current has fallen to 0 mA due to either the current sensor becoming faulty or its input lead becoming disconnected.

+4 mA to +20 mA setup characteristic

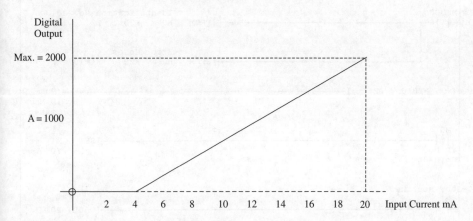

Figure 34.18

The characteristic shown in Figure 34.18 can be confirmed by writing the value H3310 to buffer memory 0.

34.22 Voltage/current specification

The specification for the FX2N-4AD program ADC1 is as follows:

1. CH1 will:
    (a) Convert a voltage input signal.
    (b) Have an offset of 2 V.
    (c) Have a gain value of 6 V.

(d) Produce a maximum digital output of 2000, when Vin is 10 V.
(e) Produce a digital output of 1500, when Vin is 8 V. Reference the voltage example on page 353.
2. CH2 will:
   (a) Convert a current input signal.
   (b) Have an offset of 4 mA.
   (c) Have a gain value of 12 mA.
   (d) Produce a maximum digital output of 2000, when Iin is 20 mA.
   (e) Produce a digital output of 375, when Iin is 7 mA (refer the current example on page 354).
3. CH2 and CH3 will be OFF.

34.23 Ladder diagram – ADC1

Main program

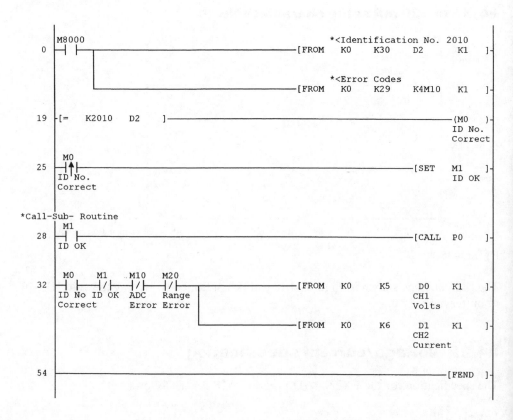

Sub-routine

```
*Sub-Routine
*EEPROM Write Delay Timers
      P0    M8000                                                            K5
      55    ──┤ ├─────────────────────────────────────────────────────────(T0  )─
                                                                             K10
            ──────────────────────────────────────────────────────────────(T1  )─
                                                                             K15
            ──────────────────────────────────────────────────────────────(T2  )─
                                                                             K20
            ──────────────────────────────────────────────────────────────(T3  )─

*CH1-Voltage -10V to +10V
*CH2-Current -20mA to +20mA
             T0
      69    ──┤ ├─────────────────────────────[TOP   K0    K0    H3320   K1  ]─

                                              [TOP   K0    K21   K1      K1  ]─

                                              [TOP   K0    K22   K0      K1  ]─

*CH1 Offset +2 Volts
     Gain   +6 Volts
             T1
      97    ──┤ ├─────────────────────────────[TOP   K0    K23   K2000   K1  ]─

                                              [TOP   K0    K24   K6000   K1  ]─

                                              [TOP   K0    K22   K3      K1  ]─

*CH2 Offset +4mA
     Gain   +12mA
             T2
     125    ──┤ ├─────────────────────────────[TOP   K0    K23   K4000   K1  ]─

                                              [TOP   K0    K24   K12000  K1  ]─

                                              [TOP   K0    K22   K12     K1  ]─

             T3
     153    ──┤ ├────────────────────────────────────────────────[RST    M1  ]─
                                                                        ID OK
*Return to Main Program
     155    ─────────────────────────────────────────────────────────── [SRET]─

     156    ─────────────────────────────────────────────────────────── [END ]─
```

34.24 Principle of operation – ADC1

1. Line 0
   (a) Buffer memory location, i.e. BFM 30, contains the identification code for the FX2N-4AD, which is 2010. This number is transferred to data register D2.
   (b) The contents of BFM 29, i.e. error status, is transferred to individual bits from M0 to M12 (refer page 359).

b0	transferred to	M10	ON	Bits b1 to b3 ON.
b1	transferred to	M11	ON	Offset/gain error.
b2	transferred to	M12	ON	Power supply abnormality.
b3	transferred to	M13	ON	Hardware failure.
b10	transferred to	M20	ON	Digital range error.
b11	transferred to	M21	ON	Averaging error.
b12	transferred to	M22	ON	Offset/gain adjust inhibit.

2. Line 19
   (a) The instruction -[=K2010 D2]- is used to check that there is an FX2N-4AD unit in block 0.
   (b) If the correct identification code has been transferred to D2 then M0 turns ON.
3. Lines 25 and 28
   (a) The momentary operation of the M0 contacts sets M1 ON.
   (b) The closing of the M1 contacts will now enable the sub-Routine to be Called. It is the sub-routine instructions, which set up the offset and gain values of the ADC.
   (c) At the end of the sub-routine, an instruction will reset M1 – Line 153.
   (d) This method of using M0 and M1 ensures that the remaining main program instructions, which read the ADC Digital values cannot be executed, until the sub-routine initialisation instructions have been executed.
   (e) Also since M0 is pulsed, then the sub-routine will only be called once, each time the CPU is set to RUN. This ensures that the EEPROM is not written to, on each scan of the program (refer page 360).
4. Line 32
   (a) Provided that the control signals M0, M1, M10 and M20 are OK, information can be read back from the ADC to the FX2N CPU.
   (b) The instruction -[FROM K0 K5 D0 K1]- transfers the contents of BFM5 to D0.
   (c) The instruction -[FROM K0 K6 D1 K1]- transfers the contents of BFM6 to D1.
   (d) This enables the digital outputs from CH1 and CH2 to be monitored.
      D0 monitors the digital output from CH1 ($-10\,V$ to $+10\,V$).
      D1 monitors the digital output from CH2 ($-20\,mA$ to $+20\,mA$).

Sub-routine instructions

The sub-routine instructions setup the initialisation of the FX2N-4AD, i.e.

(a) Channel 1 to convert input voltages in the range $+10\,V$ to $-10\,V$.
(b) Channel 2 to convert input currents in the range $-20\,mA$ to $+20\,mA$.

(c) Channel 1 will have an offset voltage of +2 V and a gain value of +6 V.
(d) Channel 2 will have an offset current of +4 mA and a gain value of +12 mA.

1. Line 55
   (a) When the sub-routine is called, the four timers – T0, T1, T2 and T3 – are energised.
   (b) These timers have time delays of 0.5 sec, 1.0 sec, 1.5 sec and 2 sec respectively.
   (c) Hence the time difference between each timer operating is 0.5 sec.
   (d) This time difference is required when writing to the EEPROM memory locations (refer page 360).
2. Line 69
   (a) After a delay of 0.5 sec from the start of the sub-routine, the channel initialisation value H3320 is now sent to BFM 0.
   (b) This means that:
   CH1 = 0, this enables CH1 to convert a −10 V to +10 V input signal.
   CH2 = 2, this enables CH2 to convert a −20 mA to +20 mA input signal.
   CH3 = 3, therefore CH3 is OFF.
   CH4 = 3, therefore CH4 is also OFF.
   (c) The instruction -[TOP K0 K21 K1 K1]- is used to permit changes to be written to BFMs 22 and 23.
   The permit data being: b1 = 0 and b0 = 1.
   (d) The instruction -[TOP K0 K22 K0 K1]- is the instruction which enables the channel initialisation value of H3320 to be transferred to the EEPROM.
3. Line 97
   (a) After a delay of 1 second the Channel 1, Offset and Gain voltages will be stored into BFM 23 and BFM 24 respectively.
   (b) Offset voltage = 2 V (K2000)
   Gain voltage = 6 V (K6000).
   (c) The values will then be transferred into the channel 1 EEPROM memory locations, when the instruction -[TOP K0 K22 K3 K1]- is executed.
   (d) The data value K3 = 0000 0011.
   This will set both G1 and O1 in BFM 22 to '1', hence enabling the offset and gain values to be transferred to the EEPROM (refer page 360).
4. Line 125
   (a) After a delay of 1.5 sec the channel 2, offset and gain currents will now be stored into the same buffer memories BFM 23 and BFM 24 respectively.
   (b) Offset current = 4 mA (K4000)
   Gain current = 12 mA (K12000).
   (e) When the instruction -[TOP K0 K22 K12 K1]- is executed, the data value K12, i.e. 0000 1100 will now set both G2 and O2 to '1'. This will enable the offset and gain values for channel 2 to be transferred to the channel 2 EEPROM memory locations.
5. Line 153
   (a) After a delay of 2 sec, the contacts of T3 will close and reset M1.
   (b) The instruction -[RST M1]- ensures that the sub-routine cannot be recalled, until the RUN switch has been turned OFF and then back ON – Line 28.
6. Line 155
   The instruction -[SRET]- causes the program to return to the main program.

34.25 Practical – analogue to digital conversion

Hardware diagram

Connect up the hardware diagram as shown in Figure 34.19 but initially leave the input to Channel 2 disconnected.

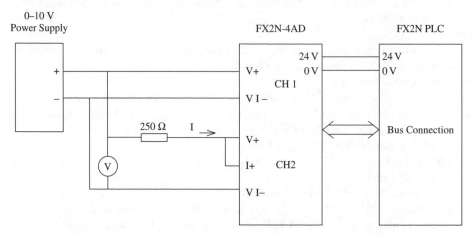

Figure 34.19

Note: The input resistance between the I^+ and V I– terminals is 250 Ω, therefore with a 250 Ω externally connected resistor also of 250 Ω, the total circuit resistance will be 500 Ω.

Hence $I = \dfrac{V \text{ supply}}{500 \, \Omega}$ Amps

Voltage (V)	Current (mA)
0	0
1	2
2	4
3	6
3.5	7*
5	10
6	12
7	14
8*	16
9	18
10	20

*$I = 7$ mA, current example – Page 353.
*$V = 8$ V, voltage example – Page 352.

Analogue to digital conversion FX2N-4AD 367

34.26 ADC results

1. Execute the program ADC1.
2. Monitoring the devices shown below, using entry data monitoring (see Figure 34.20).
   (a) D0
   (b) D1
   (c) D2
   (d) M0
   (e) M1.
3. This will enable an initial check, to ensure the ADC unit is working correctly.
4. The value in D0 is obtained with an input voltage of 8 V into channel 1. The value in D1 is obtained with the input into channel 2 disconnected.

Device	ON/OFF/Current	Setting value	Connect	Coil	Device comment	
D0	1499				CH1	Volts
D1	-498				CH2	Current
D2	2010				ID No	
M0				1	ID No	Correct
M1				0	ID OK	

Figure 34.20

34.27 Monitoring using buffer memory batch

The GX-Developer software has the facility to directly monitor the values in the buffer memory of a special unit, i.e. the FX2N-4DA, without having to first transfer the BFM contents to data registers.

This facility is known as buffer memory batch.

1. Select the following:
   (a) Online
   (b) Monitor
   (c) Buffer memory batch.
2. Enter the following values, to enable the display to become as shown in Figure 34.21.
   (a) Module start address   0.
   (b) Buffer memory address   0.
3. Select Start monitor.
4. Refer page 363, Line 69.
   The value in BFM 0 is 13088.
   The equivalent hexadecimal value is 3320, which is the initialisation setup value.
   Channel 1   0–10 V.
   Channel 2   −20 mA to +20 mA.
   Channels 3 and 4   OFF.

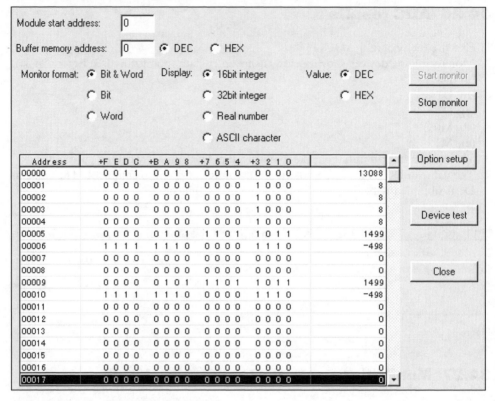

Figure 34.21

5. In BFMs 5 and 9 is the channel 1, +8 V converted value, i.e. 1499.
6. In BFMs 6 and 10 is the channel 2 open circuit current converted value, i.e. −498.
7. In BFMs 23 and 24 are the channel 2 offset and gain values respectively, i.e.
   Offset = 4000 (4 mA).
   Gain = 12000 (12 mA).

Note: These were the last values, which were transferred to BFMs 23 and 24.

34.28 Test results

(a) Ensure the circuit is connected as shown on page 366 with the input to channel 2 now connected.
(b) For channels 1 and 2, complete the results tables shown on page 369.

Channel 1 voltage results

Refer page 352, voltage example.

Vin (D0)	Digital Output	
0		
2		Offset
4		
6		Gain Value
8		Voltage Example
10		

Channel 2 current results

Refer page 353, current example.

V Supply	Iin – mA (D1)	Digital Output	
0	0		
1	2		
2	4		Offset
3	6		
3.5	7		Current Example
4	8		
5	10		
6	12		Gain Value
7	14		
8	16		
9	18		
10	20		

35

Digital to analogue conversion FX2N-4DA

35.1 Introduction

The FX2N-4DA is a 4-channel digital to analogue converter (DAC).
    Each of the four channels CH1, CH2, CH3 and CH4 are used to convert the first 12 bits of a 16 bit data register into a corresponding analogue output voltage or current.
    The output voltage range is $-10\,V$ to $+10\,V$.
    The output current ranges are $+4\,mA$ to $+20\,mA$ and $0\,mA$ to $+20\,mA$.
The DAC is used to output an analogue voltage/current to:

1. Display information, i.e.
   (a) Analogue meter
   (b) Chart recorder
   (c) Oscilloscope.
2. Control a range of output devices, i.e.
   (a) Linear actuator
   (b) Rotary valve
   (c) Speed controller – Inverter.

35.2 Voltage resolution

The voltage resolution of a DAC is the change in the analogue output voltage for a 1-digit change in the digital input (see Figure 35.1).
    The digital input is stored in the first 12 bits of a 16-bit data register.
Hence the range of numbers that can be stored is from -2^{11} to $+2^{11}$, i.e. -2047 to $+2047$.
A 1-digit change in the digital input will produce a change of $10\,V/2047$ in the output voltage.
Therefore, the voltage resolution of an 11 bit DAC $= 4.89\,mV$
However, the Mitsubishi stated voltage resolution for the FX2N-4DA $= 10\,V/2000 = 5\,mV$

Voltage characteristic

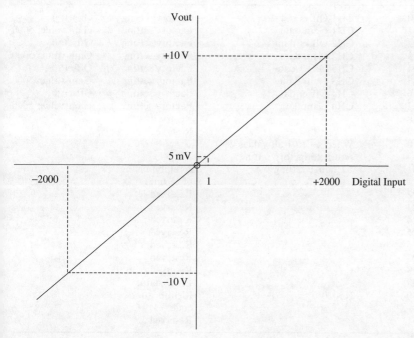

Figure 35.1

35.3 FX2N-4DA buffer memory addresses and assignments

Within the DAC is a buffer memory, BFM, which enables data transfer between the DAC and the CPU. There are a total of 32 addressable buffer memories each containing 16 bits.

	BFM		Contents	
W E	0	Output mode select	Factory setting H000	CH1–CH4 voltage output
W	1		CH1 digital input	
W	2		CH2 digital input	
W	3		CH3 digital input	
W	4		CH4 digital input	
W E	5	Data holding mode	Factory setting	H0000
W	6		Reserved	
W	7		Reserved	
W E	8	CH1–CH2 Offset/gain	Transfer to EEPROM	
W E	9	CH3–CH4 Offset/gain	Transfer to EEPROM	

(*Continued*)

	BFM		Contents	
W	10	CH1 Offset data	Factory setting	Offset: 0
W	11	CH1 Gain data	Factory setting	Gain value: 5000
W	12	CH2 Offset data	Factory setting	Offset: 0
W	13	CH2 Gain data	Factory setting	Gain value: 5000
W	14	CH3 Offset data	Factory setting	Offset: 0
W	15	CH3 Gain data	Factory setting	Gain value: 5000
W	16	CH4 Offset data	Factory setting	Offset: 0
W	17	CH4 Gain data	Factory setting	Gain value: 5000
	18		Reserved	
	19		Reserved	
W E	20	When set to 1 all values return to factory setting		
W E	21	1 Change settings	2 Inhibit change of settings	
	22		Reserved	
	23		Reserved	
	24		Reserved	
	25		Reserved	
	26		Reserved	
	27		Reserved	
	28		Reserved	
	29		Error status	
*	30	K3020	Identification code – DAC	
*	31		Reserved	

Notes:

W Write to buffer memory
E Data transfer to non-volatile EEPROM
* Read from buffer memory

35.4 Error codes – BFM 29

If an error or errors occur within the DAC, then by monitoring the contents of BFM 29 it is possible to determine what has occurred.

The instruction shown in Figure 35.2 is used to transfer the 16-bit contents of BFM 29 from the DAC to the M coils M10–M25, within the FX2N PLC.

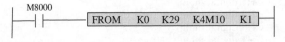

Figure 35.2

The resultant bit pattern when used with the table below will enable the error or errors to be determined.

Digital to analogue conversion FX2N-4DA

Bit		ON	OFF	M coil
b0:	Error	When any of b1 to b3 is ON	No error	M10
b1:	Offset/gain error	Offset/gain data in EEPROM is corrupted or adjustment error	Offset/gain data normal	M11
b2:	Power abnormality	DC 24 V power supply failure	Power supply normal	M12
b3:	Hardware error	D/A converter or other hardware failure	Hardware normal	M13
b10:	Range error	Digital input or analogue output value is outside the specified range	Digital output or analogue input value is inside the specified range	M20
b12:	Offset/gain adjust prohibit	BFM 21 is not set to 1 Changes inhibited	BFM 21 is set to 1 Adjustable status	M22

Note:
Bits b4–b9, and bits b11, b13, b14 and b15 are undefined.

35.5 Hardware diagram

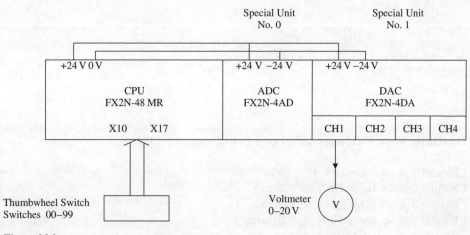

Figure 35.3

35.6 DAC special unit no. 1

By virtue of it being the second special unit within the system, the DAC block number address for all of the TO/FROM instructions will be K1.

The block number address for the ADC being K0.

35.7 Output mode select

To enable channels 1–4 to output either voltage or current, a 4-digit hexadecimal number has to be entered into buffer memory 0 (refer page 371).

The least significant character controls channel 1 whereas the most significant figure controls channel 4. Each channel can be individually set to output either voltage or current.

Output mode settings

Data value 0: Pre-set range (−10 V to +10 V).
Data value 1: Pre-set range (+4 mA to +20 mA).
Data value 2: Pre-set range (0 mA to +20 mA).

Example

Let Channels 1–4 be set up as follows (see Figure 35.4):

(a) Channel 1 set for voltage (−10 V to +10 V).
(b) Channels 2 and 3 set for current (+4 mA to +20 mA).
(c) Channel 4 set for current (0 mA to +20 mA).

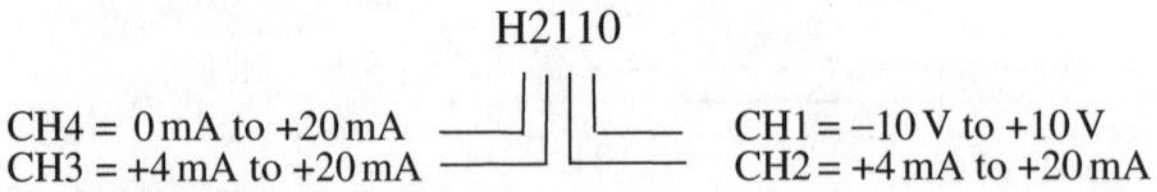

```
                        H2110
CH4 = 0 mA to +20 mA  ─┐ │ │ ┌─  CH1 = −10 V to +10 V
CH3 = +4 mA to +20 mA ──┘ └──    CH2 = +4 mA to +20 mA
```

Figure 35.4

Therefore, when H2110 is entered into buffer memory 0 then:

1. Channel 1 is set for −10 V to +10 V.
2. Channel 2 is set for +4 mA to +20 mA.
3. Channel 3 is set for +4 mA to +20 mA.
4. Channel 4 is set for +0 mA to +20 mA.

35.8 Ladder diagram – DAC1

This project enables a 2-digit number in the range 0–99 to be entered from the thumbwheel switches and then scaled up by a factor of 20.

The digital value is then supplied to channel 1 of the DAC, which will output a corresponding dc voltage in the range 0–9.9 V.

Digital to analogue conversion FX2N-4DA

```
* Digital to Analogue Converter
* DAC1
       M8002                               *<Channels 1-4 Voltage Output
   0   ─┤├─────────────────────────────[TO    K1      K0      H0      K1    ]─
       Run  │
       Pulse│
            │                              *<Inhibit Offset/Gain Adj.
            └─────────────────────────[TO    K1      K21     K2      K1    ]─

       M8000                               *<Obtain DAC ID Number
  19   ─┤├─────────────────────────────[FROM  K1      K30     D4      K1    ]─
       Run

  29   ─[ = K3020  D4 ]─────────────────────────────────────────( M0       )─
                Check                                             ON-ID
                ID No                                             Correct

       M0                                  *<Error Data-Transfer M10-M25
  35   ─┤├─────────────────────────────[FROM  K1      K29     K4M10   K1    ]─
       ON-ID
       Correct

       M8000                               *<Input Thumbwheel Switch Value
  45   ─┤├─────────────────────────────[BIN           K2X10   D0    ]─
       Run  │
            │                              *<Multiply-Obtain Max V-Output
            └─────────────────────────[MUL           D0      K20     D1    ]─

       M0      M10     M20     M22         *<Send Digital Value to CH1 DAC
  58   ─┤├─────┤/├─────┤/├─────┤├──────[TO    K1      K1      D1      K1    ]─
       ON-ID   Error   Range   Inhibit
       Correct BFM29   Error   Changes
               b0-b3   b10     BFM21=2

  71                                                                  ─[ END ]─
```

35.9 Principle of operation – DAC

1. Line 0
   (a) On switching the CPU to RUN, the contacts of M8002 will close for one scan period. This enables the instructions -[TO K1 K0 H0 K1]- and -[TO K1 K21 K2 K1]- to be executed just once, each time the CPU is switched to RUN.

(b) The instruction -[TO K1 K0 H0 K1]- enables CH1–CH4 to be set for voltage outputs (refer page 374).
  (i) TO
   Every time the input contacts are closed, a value (i.e. H0000) is written from the CPU to the special unit (i.e. the DAC) which is located at position 1.
  (ii) K1
   This defines the block number address of the DAC.
  (iii) K0
   Buffer memory 0 (BFM 0) stores the DAC output mode.
  (iv) H0 (H0000)
   This is the output mode value required to enable the four channels, i.e. CH1–CH4 to output voltage.
  (v) K1
   The data value is only written to BFM 0.
(c) The instruction -[TO K1 K21 K2 K1]- writes the value 2 to BFM 21, which prevents any changes being made to the factory settings for offset and gain for all the four channels.

2. Line 19
   (a) The instruction -[FROM K1 K30 D4 K1]- enables the identification number for the special unit located at position 1, to be obtained from BFM 30.
   (b) As the DAC is in position 1, i.e. K1, its identification number K3020 will be transferred to D4 within the CPU.

3. Line 29
   (a) The instruction -[= K3020 D4]- compares the contents of D4 with the identification code for an FX2N-4DA, which is K3020.
   (b) As the FX2N-4DA is in position 1, the comparison will be TRUE and M0 will turn ON.

4. Line 35
   The instruction -[FROM K1 K29 K4M10 K1]- is used to transfer the 16-bit contents of BFM 29, the error status information (refer page 372) to individual memory locations starting at M10, i.e.

   $$\left.\begin{array}{l} \text{bit 0 to M10} \\ \vdots \\ \text{bit 12 to M22} \end{array}\right\} \text{Error codes b0–b12}$$

   $$\left.\begin{array}{l} \text{bit 13 to M23} \\ \text{bit 14 to M24} \\ \text{bit 15 to M25} \end{array}\right\} \text{Not used}$$

5. Line 45
   (a) On obtaining the correct identification no., the first two digits of the thumbwheel switch, which are connected to X10–X17, will be transferred to D0.

(b) It can be seen from the DAC voltage characteristics on page 371 that to obtain a maximum output voltage of 10 V, the digital input has to be 2000.
(c) Therefore since the maximum value from the 2-digit thumbwheel switches is only 99, this value has to be multiplied by 20 to give a digital value of 1980. This will enable an output voltage of 9.9 V to be produced from the DAC.
(d) The result of the multiplication is stored in D1 and D2 but since the maximum multiplied value will not exceed 32767, the value to be transferred to the DAC will always be in D1.

6. Line 58
   (a) Before the digital output value can be transferred to the DAC, the following inputs must be present.

   (i) M0 ON   Identification No. OK. The DAC is located in position 1.
   (ii) M10 OFF   Bit 0 OFF   b1 – b2 = 0 No error.
                  Bit 1   Offset/gain values in the EEPROM are OK.
                  Bit 2   Power supply OK.
                  Bit 3   Hardware OK.
   (iii) M20 OFF Digital input and analogue output are in range.
   (iv) M22 ON   BFM 21 set to 2. Changes to offset/gain settings inhibited.

   (b) The instruction -[TO K1 K1 D1 K1]- transfers the contents of D1 to the digital input of CH1, where it can be converted to an equivalent analogue output voltage.

35.10 Practical – digital to analogue conversion

Carry out the following practical exercise.

1. Connect up the system as shown in Figure 35.3.
2. Enter and execute the program DAC1.
3. Complete the table below:
   (a) Input the following digital values from the thumbwheel switches.
   (b) Measure the corresponding analogue output voltage from the FX2N-4DA.

Digital Input Thumbwheel Switches	Analogue Output Calculated (V)	Analogue Output Measured (V)
0	0	
10	1.0	
20	2.0	
30	3.0	
40	4.0	
50	5.0	
60	6.0	
70	7.0	
80	8.0	
90	9.0	
99	9.9	

36

Assignments

The assessment of BTEC Units now requires that a large proportion of the assessment use meaningful assignments. For engineering courses, this requires assignments to be based on carefully thought out industrial applications and scenarios.

With this in mind, it is hoped that the following examples can be used by my ex-colleagues in Further and Higher Education to produce meaningful and stimulating assignments for their students.

For those who are not students on a PLC engineering course, it is hoped that they can find the time to try and produce effective solutions.

Nothing can best prepare someone to produce good working PLC programs, than 'having a go' themselves once they have acquired the basic skills and knowledge, which it is hoped this book has done.

1. Using memory coils M8015–M8016 and data registers D8013–D8015 produce a 24-hour clock, which can be tested using ladder logic tester.
2. Modify INTLK1 (refer page 80) so that it could be used in a quiz in which the first contestant, who wishes to answer the question, can press the button and bring on just that particular output light. The other contestants' outputs will be inhibited, until a reset button is operated.
3. **TRAF2**
   Traffic lights are temporarily installed to control single lane access along a stretch of road where new drainage pipes are being laid (see Figure 36.1).

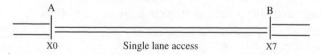

Figure 36.1

Two sets of traffic lights are installed at either end, i.e. at points A and B.

Sequence of operation

a. On turning the PLC ON, A is set to green and B is set to red.
b. Leave the lights at this setting for 30 sec.
c. Check that cars are present at point B, i.e. check Input X7.
d. If no cars are there at B, then leave the lights, i.e. A green, B red.

e. If there are cars at B, then turn both A and B to red for 20 sec.
f. This gives time for any car travelling from A to B to clear the single lane.
g. After the 20 sec, leave A at red and set B to green.
h. Leave the lights at this setting for 30 sec.
i. Check that cars are present at point A, i.e. check input X0.
j. If there are no cars at A, then leave the lights, i.e. A Red, B green.
k. If there are cars at A, then turn both A and B to RED for 20 sec.
l. This gives time for any car travelling from B to A to clear the single lane.
m. After the 20 sec, leave B at red and set A to green.
n. Repeat the sequence as from item b.

From the description of the sequence of operations and the associated flowchart (see Figure 36.2) produce a ladder diagram, which can be tested using a PLC or ladder logic tester.

Flowchart TRAF2

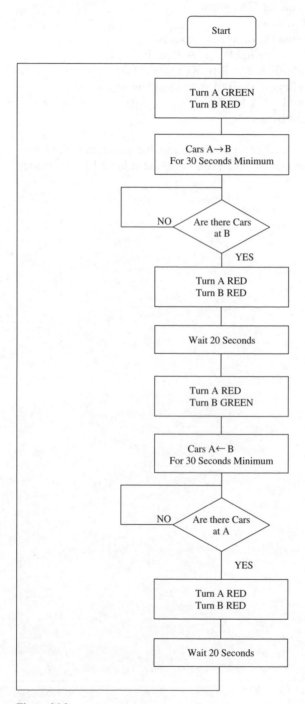

Figure 36.2

4. **2-handed safety system**
   Warning: The following project must only be used to simulate the operation of a 2-handed safety system. Under *no circumstances* must it be used as the safety circuit for an actual control system (refer page 111).
   All emergency circuits such as emergency stop buttons and safety guard switches must be hardwired and *not* depend on software, i.e. PLCs or electronic logic gates.

Assignment

Produce a PLC program, which simulates the operation of a 2-handed safety circuit, i.e. the PILTZ PNOZ e2.1P, based on the following description:
   a. When operating metalworking presses, it is essential that the operator's hands are clear of the press.
   b. Hence to start the press, two start buttons, which are outside the press and which are also an arm's length distance apart, are pressed within 0.5 sec of one another by the operator.
   c. Should one of the buttons develop a fault, i.e. a short circuit, then the press cannot start, as both buttons have to be operated within the 0.5 sec margin.
   d. Use inputs X0 and X1 as the two start buttons and input X7 as the press cycle complete signal.
   e. Use the output Y0 to signal the start of the press cycle.

5. The specification for the drive to a stepper motor is as follows:
   a. It is to be driven by 1-second pulses from the output Y0.
   b. The number of pulses, which are to be applied to the motor is obtained from the thumbwheel inputs X10–X17.
   c. If the input = 1, then one pulse is applied to the stepper motor. If the input = 99, then 99 pulses are applied to the motor.
   d. Owing to time delays in the rise and fall of the motor current caused by the resistance and inductance of the motor windings, it is essential that the Y0 signal must be ON /OFF for the full 0.5 sec.

Output Y0 – waveform

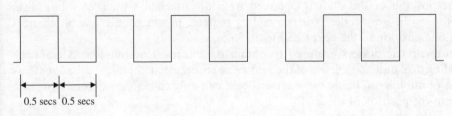

Figure 36.3

Hint: Investigate the instruction PLSY.

6. In a kitchen furniture company the kitchen units are assembled and then transported by conveyors to a machine, where they are wrapped and boxed.

   When an operator at the assembly area pushes a button X0 then if there is no unit on conveyor 2, the unit will leave conveyor 1 and move onto conveyor 2.

   Similarly if there is no unit on conveyor 3, the unit will travel from conveyor 2 onto conveyor 3 and wait at the end of conveyor 3 until called into the wrapping machine by the operation of input X4.

   Photocell X3 is used to stop conveyor 4 when the end of the unit clears the photocell.

   The conveyors when used as described above are known as accumulating conveyors, i.e. they accumulate the units onto the conveyors.

   Produce a ladder diagram, which will control the operation of the conveyors shown in Figure 36.4.

Block diagram – accumulating conveyor system

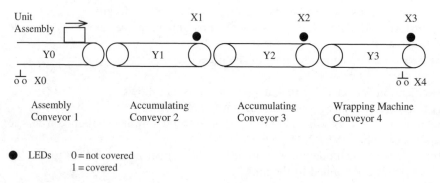

Figure 36.4

7. The sequential operation of a process plant is obtained by the angular position of its main drive shaft. That is at known positions of the shaft, selected outputs will be turned ON and OFF, i.e. similar to a drum controller.

   The position of the shaft is obtained from an attached 4-track encoding disk, which uses the gray code. This is a coding pattern, in which only one bit changes from one position to the next (see Figure 36.5).

   Photo-electric devices are used to obtain the electrical output signals from the encoding disk unit and the use of the gray code ensures that any slight misalignments in the positioning of the photo-electrical cells can only cause a maximum error of 1 place in the position of the shaft.

   It is necessary to decode from gray code to pure binary to determine what process operation is to be carried out next.

   The conversion process can be done using digital logic half adder circuits, but since a PLC is already being used, it would be economical if the conversion could be done instead, using a PLC program.

Simulate the encoder disk position, using the inputs X0–X3 and display the converted true binary, on the outputs Y0–Y3.

Shaft position	Gray code input				Binary output			
	X3	X2	X1	X0	Y3	Y2	Y1	Y0
0	0	0	0	0	0	0	0	0
1	0	0	0	1	0	0	0	1
2	0	0	1	1	0	0	1	0
3	0	0	1	0	0	0	1	1
4	0	1	1	0	0	1	0	0
5	0	1	1	1	0	1	0	1
6	0	1	0	1	0	1	1	0
7	0	1	0	0	0	1	1	1
8	1	1	0	0	1	0	0	0
9	1	1	0	1	1	0	0	1
10	1	1	1	1	1	0	1	0
11	1	1	1	0	1	0	1	1
12	1	0	1	0	1	1	0	0
13	1	0	1	1	1	1	0	1
14	1	0	0	1	1	1	1	0
15	1	0	0	0	1	1	1	1

4-bit gray coded disk

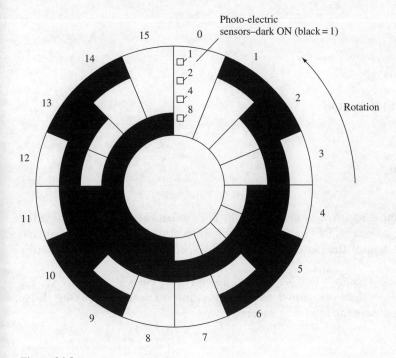

Figure 36.5

8. A mail order company has to sort customers' deliveries, according to the region where they are to be sent, i.e. North, Midlands and South.

The packages initially travel along a conveyor system where they are scanned by a bar code reader. Based on the bar code reading, one of three output solenoids will be energised to push a package into the correct bin.

Use three switches connected to inputs X2–X4 to simulate the bar code reader and connect a push button to input X5 to simulate the movement of the conveyor (see Figure 36.6).

For the sake of simplicity assume that the packages are all of the same length and that the operation of the push button X5 simulates the movement of the conveyor a distance equivalent to the length of one of the packages.

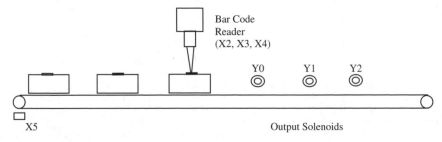

Figure 36.6

Let X0 = Start
    X1 = Stop.
Bar code simulator
    X2 = North
    X3 = Midlands
    X4 = South.
Conveyor movement simulator
    X5.
Output solenoids
    Y0 = North
    Y1 = Midlands
    Y2 = South.

9. A milling machine is required to mill a straight-line cut between two points on a flat steel plate. This is done by having the rotating milling cutter in a fixed position and the table, which clamps the plate, being driven by separate motors in the X and Y directions.

The start of the cut is at position X1, Y1 and the end of the cut is at positions X2, Y2.

The co-ordinate values are stored as floating point values in a Look-up Table starting at D0 and have the following values in mm.

Co-ordinate	Data Registers	Value (mm)
X1	D0/D1	24.5
Y1	D2/D3	30.0
X2	D4/D5	100.5
Y2	D6/D7	60.4

The speed of the motor is such that the clamping table moves in the X direction at 1000 mm/min. Produce a ladder diagram program, which will determine and display the table speed in the Y direction.

10. A fixed position plasma cutter is used to cut a circle of diameter 2 m from a flat steel plate by moving the bed to which the plate has been bolted.

    The X and Y movements of the bed are controlled by two servo-driven motors. The co-ordinates for the centre of the circle are x = 2.5 m, y = 2 m. Produce a ladder diagram program, which will enable the plate to be cut.

    Hint: Let the distance the bed moves be incremented a distance of 50 mm every 0.1 sec from its datum position (0, 0) to its maximum value and then decremented 50 mm every 0.1 sec back to the datum position. The corresponding Y movement being determined, using Pythagoras theorem.

11. An ADC card is setup as follows:
    (a) Offset = 2 V.
    (b) Gain value = 6 V.
    The ADC is to be used to monitor voltages in the range 5–7 V.

    Produce a modification to the PLC program ADC1, which will:
    (a) Turn only Y0 ON, while in the specified range.
    (b) Flash only Y6 ON/OFF at 1-second intervals, if out of range and the input has not become disconnected.
    (c) Flash only Y7 ON/OFF at 1 second intervals, should the input become disconnected.

12. It is required to turn off water flowing through a large mains pipe using an electrically operated rotary valve.

    The electrical supply to the valve is obtained from a digital to analogue converter whose output is in the range 0–10 V. The input to the converter is from a data register whose range is from 0 to 2000.

    To ensure the valve closes without any problems, the supply to operate the valve is: $V_{supply} = 10[1 - e^{-t}(1 + t)]$ i.e. a critically damped waveform.

    Produce a modification to ladder diagram DAC1, which can produce the required output voltage.

    Hint: Use a spreadsheet to obtain the values for V supply from 0 to 10 sec at intervals of 0.2 sec and transfer these values into the PLC's device memory.

Index

Acknowledgements, xvii
Address range FX2N devices, 15
Advanced programming instructions, 198–199
Analogue to digital conversion – ADC, 344–369
  ADC equations, 349–350
  buffer memory addresses, 344–345
  channel initialisation, 356–357
  count averaging, 354
  EEPROM:
    buffer memory, 360
    operational life cycles, 360
    write time delay, 360
  errors – BFM 29, 359
  examples:
    current conversion, 353–354
    voltage conversion, 352–353
  hexadecimal numbering – special units, 356
  input offset voltage, 348
  input voltage limitations, 350
  instructions:
    TO and FROM, 345, 357–359
  introduction, 344
  offset and gain programming, 360–361
  open circuit input detection, 361
  positioning of ADC unit, 355
  program/project – ADC1, 361–369
    buffer memory batch monitoring, 367–368
    hardware diagram, 366
    ladder diagram, 362–363
    practical, 366
    principle of operation, 364–365
    specification, 361–362
    test results, 368–369
  resolution:
    definition, 350
    highest possible, 351–352
    maximum input – voltage/current, 346–347
  voltage and current conversion, 345–346
  voltage gain value, 348
  wiring diagram, 355–356
Assignments, 378–385
  24-hour clock, 378
  accumulating conveyors, 382
  ADC – voltage monitor, 385
  BATCH2, 87
  DAC – rotary valve controller, 385
  gray to binary conversion, 382–383
  milling machine, 384–385
  package sorting, 384
  plasma cutter, 385
  PNEU2, 104
  quiz response, 378
  stepper motor controller, 381
  traffic lights – single lane operation, 378–380
  two-handed safety system, 381
Automatic queuing system, 332–343
  basic operation, 333–334
  description, 332
  FIFO memory stack, 333
  logical value, 333, 340
  program/project QUEUE1, 334–343
    analysis of results, 343
    input scan waveforms, 337
    instructions:
      FMOV, 334
      INCD, 334
      PLF, 336, 342
      SFRDP, 335
      SFWR, 334
    ladder diagram, 334–336
    monitoring, 343
    output truth table, 340
    principle of operation, 336–342
    testing, 342
  software diagram, 333
  system hardware, 332
Auxiliary memory coils:
  address range, 15
  bi-directional counters, 245
  description, 14
  high speed counters, 251
  interrupt, 320

Auxiliary memory coils:*(Contd)*
  list, special purpose, 171
  symbol, 14

Base unit, extension unit and extension blocks, 4
Basic PLC programs, 71–90
  BATCH1, 86–87
  BATCH2, 87
  COUNT1, 82–84
  COUNT2, 84–86
  FURN1, 74–78
  INTLK1, 78–80
  LATCH1, 81
  MC1, 87–90
  TRAF1, 71–74
Basic PLC units, 1–2
Battery backup:
  device address ranges, 15
  latch relays, 80
    program/project *see* Latch relays
Battery error *see* Diagnostic fault finding
Bi-directional counters, 244–246
  program/project – COUNT5:
    description, 244
    ladder diagram, 244
    monitoring, 246
    operating procedure, 245
    principle of operation, 245
    special M coils, M8200 – M8234, 245
Binary coded decimal (BCD), 197
  program/project – COUNT4, 202–204
    description, 202
    ladder diagram, 202
    monitoring, 203–204
      BCD display, 203
    principle of operation, 203
Binary counter, 200–202
  program/project – COUNT3:
    description, 200
    entering the program, 201
    ladder diagram, 201
    monitoring, 202
    principle of operation, 201
    system block diagram, 200
Binary numbers, 195–197
BTEC, xv, 378

Central processor unit – CPU, xxi, 1
Change of i/o address, 48–49

Change of PLC type, 162–164
  program/project – PNEU1D, 163–164
    ladder diagram, 164
Comments *see* Documentation
Compare instructions:
  inline, 199
  using M coils, 198–199
Comparison of PLC and Relays systems, 2
Configuration – hardware, 3–9
Conversion to an Instruction Program, 27–28
Converting binary numbers to decimal, 196–197
Converting MEDOC to Gx-Developer, 157–161
  description, 157
  program/project – PNEU1C:
    importing from MEDOC, 157–159
    ladder diagram, 160–161
Copying a project *see* Modifications to an existing project
Counter programs:
  COUNT1, 82
  COUNT2, 84
  COUNT3, 200
  COUNT4, 202
  COUNT5, 244
  HSC1, 253
  HSC2, 254
  HSC3, 257
  HSC4, 260
  MIXER1, 210
Counters:
  BCD, 202–204
  bi-directional, 244–246
  binary, 200–202
  high speed, 247–264
  indirect setting using a data register – MIXER1, 210–213

Data logger, 315
Data registers, 195–199
  16/32 bit, 195
  address range, 15
  advanced programming instructions, 198–199
  format, 195
  introduction to programs using data registers, 200–213
    COUNT3, 200
    COUNT4, 202

MATHS1, 205
MIXER1, 210
REV1, 206
REV1A, 209
Decade divider – HSC2 *see* High speed counters
Deleting *see* Modifications to an existing project
Diagnostic fault finding, 165–170
  battery error, 166
  CPU error, 165
  program errors, 166
  program/project – PNEU3, 167–170
    help display – program errors, 168–169
    ladder diagram, 167
    program error check, 169–170
      FLASH1, 28–29
      PNEU3, 169–170
Digital to analogue conversion – DAC, 370–377
  buffer memory addresses, 371–372
  DAC unit block address number, 373
  description, 370
  error codes – BFM 29, 372–373
  output mode select, 374
  program/project – DAC1, 373–377
    hardware diagram, 373
    instructions – TO and FROM, 376
    ladder diagram, 374–375
    practical, 377
    principle of operation, 375–377
  voltage resolution, 370–371
Display:
  change of colour, 134–135
  comments, statements and notes, 135–137
Division:
  instruction, 199
  program/project – SUB1 *see* Sub-routines
Documentation, 116–150
  comments, 117–122
    15/16 character format, 125–126
    32 character format, 128–130
    capacity, 147
    change of colour, 134–135
    description, 116
    display of comments and statements, 124–127
    display of comments, statements and notes, 135–137

downloading to a PLC, 147–148
  entering, 118–122
    input, 119–120
    M coil, 121–122
    output, 119–121
    project data list, 118
    timer, 122
  saving in the PLC, 146–148
  uploading from the PLC, 148–150
    ladder diagram – PNEU1B, 150
notes:
  description, 117
  entering output details, 130–131
printouts *see* Print
segment/note-block edit, 132
statements:
  description, 116
  embedded, 123
  entering, 123–124
  ladder diagram search using statements, 133–134
  separate, 123
Downloading *see* Serial transfer
DWR – look up table, 300

Edit menu, 47
  delete line/branch, 52
  insert line, 47
  Write mode – F2, 67
Electrical noise immunity:
  ADC, 355–356
  comparison – PLC with relay systems, 2
Emergency stop:
  requirements, 111
  safety relay – fault conditions, 114–115
  wiring diagram, 113
Encoders:
  A/B phase encoder, 248
  A/B phase waveforms, 249, 263
  accuracy, 249
  industrial, 249
  single phase – HSC3 *see* High speed counters
End instruction, 27
Entry ladder monitoring, 151–156
  copy and paste, 153–155
  deleting, 156
  description, 151

Entry ladder monitoring,*(Contd)*
　display, 155
　ladder diagrams:
　　entry ladder monitoring, 155
　　PNEU1, 152
　monitoring, 155–156
　　display, 156
　　start – F3, 155
　principle of operation, 153–156
Errors *see* Diagnostic fault finding
Excel spreadsheets:
　copy and paste to Gx –Developer, 309
　entering the RECIPE1 look up table, 309
Executing a project – RUN, 56
Extension:
　blocks, 4
　units, 4

Fault finding *see* Diagnostic fault finding
FEND instruction:
　interrupt, 319
　sub-routine, 311
Files – importing a MEDOC file into Gx-Developer, 157
Find *see* Program search
Floating point numbers, 265–274
　accurate integer result, 272
　description, 265
　format, 268
　instructions:
　　DEADD, 266
　　DEBCD, 266
　　DEBIN, 266
　　DEMUL, 266
　　DINT, 266
　　FLT, 265
　　INT, 266
　ladder logic tester, 273
　number range, 265
　programs/projects:
　　FLT1, 266–270
　　　conversion – binary to floating point, 269
　　　ladder diagram, 267
　　　monitoring – floating point numbers, 270
　　　monitoring – integer numbers, 267–268

　　　principle of operation, 266–267
　　　storing floating point numbers, 266
　　FLT2 – area of a circle, 270–273
　　　ladder diagram, 271
　　　monitored results, 273
　　　principle of operation, 272
Forced input/output, 101–104
　description, 101
　execution history, 103–104
　forcing inputs, 101–102
　forcing outputs, 103
　forcing with the PLC in RUN, 101
　ladder logic tester, 216–217
Free line drawing, 105–110
　description, 105
　program/project – PNEU1A, 105–110
　　connecting an additional output, 108
　　deleting, 109–110
　　entering the ladder diagram, 106–107
　　ladder diagram, 107
　　modified ladder diagram, 108
Function keys:
　F2 – write mode (stop monitor), 67
　F3 – start monitor, 66
　F4 – convert to an instruction program, 28
Furnace temperature controller, 74–78
　description, 74
　program/project – FURN1, 75–78
　　algorithm, 76
　　block diagram, 75
　　ladder diagram, 77
　　principle of operation, 77–78
　　safety procedure, 76
　　temperature characteristic, 75
　　temperature sensors, 76
FX2N PLC:
　address range, 15
　basic operation, 15–17
　　block diagram, 16
　　ladder diagram, 15
　　principle of operation, 17

Gray code assignment, 382–383
Guard switches – safety, 111
Gx-Developer:
　preface, xxv
　software, 3
　start up procedure, 18–23

Header *see* Print
Help display *see* Diagnostic
    fault finding
High speed counters, 247–264
  description, 247
  encoders:
    A/B phase disk, 248
    A/B phase waveforms, 249, 263
    accuracy, 249
    industrial , 249
  FX range, 249–250
  input assignment, 250–251
  instructions:
    DHSCR, 255
    DHSCS, 259
    DHSZ, 255
  maximum frequency – total, 252
  programs/projects:
    HSC1 – up/down, 253–254
      block diagram, 253
      ladder diagram, 253
      principle of operation, 254
      switch de-bounce capacitor, 254
    HSC2 – decade divider, 254–257
      description, 254–255
      ladder diagram, 255
      monitoring, 257
      principle of operation, 257
      switch de-bounce capacitor, 255
      waveforms, 256
    HSC3 – motor controller,
      257–260
      description, 257–258
      ladder diagram, 259
      motor speed/brake waveforms,
        258
      principle of operation, 260
      sequence of operation, 258
    HSC4 – A/B phase, 260–264
      description, 260–261
      ladder diagram, 262
      principle of operation,
        263–264
      waveforms, 263
      wiring diagram, 261
  special memory coils – up/down counting,
    251
  types of counters:
    A/B phase, 248
    single phase, 247–248
    two phase bi-directional, 248

Index registers, 293–297
  BREW1 *see* Recipe
    application – BREW1
  description, 293
  instructions, 293
  offset, 293
  program/project – INDEX1, 294–297
    description, 294
    ladder diagram, 295
    look up table, 294
    monitoring, 296–297
    principle of operation, 295–296
    system block diagram, 294
  registers:
    V0 – V7, 293
    Z0 – Z7, 293
Inputs, 8–14
  AC inputs, 8
  address range, 15
  devices, 8
  ladder diagram symbol, 13
  processing, 17
  proximity sensors, 12–13
    PNP/NPN, 12
    S/S terminal configuration, 13
  source/sink, 10–12
    block diagrams, 11, 12
    description, 10
    direction of current flow:
      sink, 12
      source, 11
    S/S terminal connection, 11, 12
Insertion of contacts and rungs *see*
  Modifications to an existing project
Instruction:
  advanced programming list, 198–199
  converting to an instruction
    program, 27
  program/project – FLASH1, 30–31
    explanation, 30–31
    instruction program, 30
    ladder diagram, 30
  programming, 29–31
  scan and execution – BREW1, 291
Interlock circuit, 78–80
  description, 78
  program/project – INTLK1, 79–80
    block diagram, 79
    ladder diagram, 80
    principle of operation, 80
    task, 79

Interrupts, 315–323
    application, 315–316
    description, 315
    enable/disable – M Coils, 320
    pointer, 322
    program/project – INT1, 316–323
        description, 316
        instructions:
            ABSD – absolute drum, 318
            EI – enable interrupt, 318
            IRET – return from ISR, 319
        ladder diagram, 318–319
        monitoring, 322–323
        principle of operation, 319–322
        sequence of operation, 316
        service routine – ISR, 315, 322
        switch bounce prevention, 321–322
        waveforms, 317
Inverter drive, 299

Labels:
    interrupt pointer, 322
    sub-routines, 311, 313
Ladder diagrams:
    entering/producing see Producing a ladder diagram
    introduction, 1
    key press numbers, 22
    symbols, 13–14
Ladder Logic Tester – LLT, 214–243
    automatic monitor mode, 216
    device memory monitor – LLT1, 217–225
        forcing an input, 217–220
        ladder diagram, 218
        monitoring data registers, 219–222
        timing charts, 222–225
            introduction, 222
            procedure, 222–224
            waveforms – obtaining and display, 224–225
    floating point, 273
    input simulation methods, 216
    input/output settings, 225–243
        description, 225–226
        LLT2 modification – inline comparison and reset, 238–239
            description, 238
            ladder diagram, 238
            monitoring, 239
            procedure, 238

program/project – LLT2, 225–239
    entering conditions and settings, 228–232
    executing the i/o system, 232–234
    ladder diagram, 226
    procedure, 226–228
    resetting a data register, 234–237
    saving LLT2.IOS, 231
installing Gx-Simulator, 214
introduction, 214
program/project – FLASH1, 214–217
    forcing an input, 216–217
    ladder diagram, 214
    program execution/procedure, 214–217
simulating PNEU1, 240–243
    ladder diagram, 240
    monitoring, 242–243
    procedure – i/o system settings, 241–242
Latch relays, 80–81
    description, 80
    memory range, 80
    program/project – LATCH1:
        ladder diagram, 81
        principle of operation, 81
Line/Step numbers, 24, 32
Look up tables see Recipe application – BREW1; Index register – INDEX1

M coils see Auxiliary memory coils
Main menu, 18
Master control:
    instructions:
        MC, 88
        MCR, 88
    nesting level, 275–279
        description, 275
        flow diagram, 275
        program/project – MC2, 276–278
            ladder diagram, 276
            monitoring, 278
            principle of operation, 277
    PNEU1, 94
    program/project – MC1, 87–90
        clock signal M8013 see special M Coils
        description, 87–88
        instruction – ALTP, 88MEDOC, xv, 157
        ladder diagram, 88

output waveforms, 90
principle of operation, 88–90
MEDOC, xv, 157
Memory capacity, 147
Mitsubishi Ltd, xv
Modifications to an existing program, 40–54
　copying a project prior to, 40–41
　programs/projects:
　　FLASH2 – insert, 42–49
　　　branch, 45–46
　　　change of i/o address, 48–49
　　　contact, 43–45
　　　line/rung, 47–48
　　　modification details, 42
　　FLASH3 – delete, 50–54
　　　branch, 52
　　　contact, 50–51
　　　modification details, 50
　　　multiple lines/rungs, 53–54
　　　single line/rung, 53
Monitoring:
　buffer memory batch, 367–368
　combined ladder and entry data, 70
　description, 66
　device batch, 171–174
　　inputs, 174
　　M coils, 171–172
　　option set up, 173
　entry data, 67–69
　　description, 67
　　monitor, 69
　　register devices, 68–69
　entry ladder see Entry ladder
　　monitoring
　floating point numbers, 270
　ladder diagram – FLASH1, 66–67
　　display, 67
　　start monitor – F3, 66
　ladder logic tester:
　　device memory, 217–220
　　i/o system, 232–243
　　ladder diagram, 216
　　timing chart waveforms, 222–225
　trace waveforms, 190–194
Multiplication:
　instruction, 199
　programs/projects:
　　BREW1 – index registers, 305
　　FLT2 – area of a circle, 271
　　MATHS1, 205
　　REV1 – RPM counter, 207

Notes see Documentation
Number representation:
　binary coded decimal – BCD, 197
　　program/project – COUNT4, 202–204
　　　ladder diagram, 202
　　　monitoring, 203–204
　　　principle of operation, 203
　binary/decimal, 195–197
　　program/project – COUNT3, 200–202
　　　description, 200
　　　entering the program, 201
　　　ladder diagram, 201
　　　monitoring, 202
　　　principle of operation, 201
　　　system block diagram, 200
　hexadecimal, 174, 356
　octal – i/o, 15, 174, 356

Octal see Number representation
On-line programming, 84–86
　description, 84
　program/project – COUNT2, 84–86
　　ladder diagram, 85
　　on-line change, 86
Opening a new project see Start-up
　procedure
Outputs:
　address range, 15
　analogue, 370
　ladder diagram symbol, 14
　processing, 17
　relay output wiring diagram – PNEU1, 99
　relay, transistor, triac, 9
　turn off using M8034, 88–89, 171

Part number, 7
Piltz Automation, xix, 111
Planned preventative maintenance
　schedules – PPMs, 316
Pneumatic panel:
　operation:
　　PNEU1, 98
　　STEP_CNTR1, 331
　pneumatic drawing – PNEU1, 99
　wiring diagram connections, 99–100
Preface, xv

394 *Index*

Print, 137–145
  header, 138–139
  multiple printing, 141–145
    ladder, 145
    parameters, 143–144
    title page, 142
  page numbers, 139–141
  title page, 138–139
Producing a ladder diagram, 24–29
  program/project – FLASH1:
    conversion to an instruction
       program – F4, 27–28
    entering, 25–27
    ladder diagram, 24
    principle of operation, 24–25
    program error check, 28–29
    saving, 28
Program search, 31–39
  contact/coil, 34–35
  cross-reference, 36–37
    jump option, 37
    list, 36
  device, 33–34
  instruction, 35–36
  list of used devices, 38–39
  step number, 32–33
Program/project:
  copying, 40–41
  downloading/write, 55–56
  name, 19
  opening, 19
  saving, 28
  search *see* Program search
  start up *see* GX – Developer start-up
     procedure
  uploading/read, 63–64
  verification, 62–63
Programmable Logic Controllers – PLC's:
  address range, 15
  basic operation, 15–17
    block diagram, 16
    ladder diagram, 15
    principle of operation, 17
  basic units:
    CPU, 1–2
    input, 2
    output, 2
    power supply, 2
  comparison with relay
     systems, 2
  glossary, xxii

inputs, 8
introduction, 1
outputs, 9
preface, xv
software requirements, 2–3
Programs/Projects:
ADC1, 361
BATCH1, 86
BATCH2, 87
BREW1, 298
COUNT1, 82
COUNT2, 84
COUNT3, 200
COUNT4, 202
COUNT5, 244
DAC1, 374
FLASH1, 24, 214
FLASH2, 40
FLASH3, 50
FLASH4, 63
FLT1, 266
FLT2, 270
FURN1, 74
HSC1, 253
HSC2, 255
HSC3, 257
HSC4, 260
INDEX1, 294
INTLK1, 78
LATCH1, 81
LLT1, 217
LLT2, 226
LLT2 – modification, 238
MATHS1, 205
MC1, 87
MC2, 276
MIXER1, 210
PNEU1, 93, 151, 240
PNEU1A, 105
PNEU1B, 148
PNEU1C, 157
PNEU1D, 163
PNEU2, 104
PNEU3, 166
PNEU4, 175, 181
QUEUE1, 334
REV1, 206
REV1A, 209
ROTARY1, 285
SHIFT1, 280
STEP_CNTR1, 325

SUB1, 312
TRAF1, 71
Project data list:
   description, 22–23
   device memory – lookup table, 300
   entering comments, 118
   icon, 23
   saving comments in the PLC, 146

Queuing system *see* Automatic
   queuing system

Recipe application, 298–309
   description, 298
   look up table, 299–300
   program/project – BREW1, 300–309
      Excel spreadsheet, 309
      instructions:
         BMOV, 305
         ENCO, 305
      ladder diagram, 305
      lookup table:
         device range, 303–304
         downloading, 303–305
         entering values (DWR), 300–302
         monitoring, 305
         project data list, 300
      monitoring, 307–309
      principle of operation, 306–307
   sequence of operation, 299
   system diagram, 299
Registers *see* Data registers
Relays:
   comparison with PLC's, 2
   output wiring diagram – PNEU1, 100
   protection diodes, 100
   safety *see* Safety
Resources, xix
Return from:
   interrupts IRET, 319
   sub-routines SRET, 311–312
Risk assessment procedure, 111
Rotary index table application, 282–292
   description, 282
   program/project – ROTARY1,
      283–292
      index table system-plan view, 282
      instruction scan and execution,
         291–292

      ladder diagram, 285–286
      monitoring procedures, 289–291
      principle of operation, 287–289
      shift register layout, 284
      system requirements, 283
RPM counter, 206–210
   description, 206
   program/project – REV1, 207–209
      ladder diagram, 207
      monitoring, 208–209
      principle of operation, 207–208
   REV1A, 209–210
Run/stop, 55, 56

Safety, 111–115
   category 2 fault condition, 112
   emergency stop circuit – PNEU1, 113
   emergency stop requirements, 111
   European machinery standards, 111
   fault conditions, 114–115
   guard switches, 111
   Piltz safety relays, 111
   risk assessment procedure, 111
   safety relay specification, 112
   system start up check, 115
   two handed safety system – assignment, 381
Saving:
   a project with a different filename
      i.e. copying, 40–41
   an existing project, 28
   comments in the PLC *see*
      Documentation
Search *see* Program search
Sequence function chart (SFC):
   PNEU1, 92–93
   PNEU4, 176
Serial number, 7
Serial Transfer, 55–65
   change of communications port, 60–62
   COM port – saving the setting, 61
   connection diagram, 55
   connection set up menu, 61
   downloading/write:
      comments, 146–148
      look up table, 303–304
      project, 55–56
   PCMCIA card, 60
   reducing number of steps transferred,
      57–58
   SC-O9 converter, 55

Serial Transfer,*(Contd)*
  system image – connection set up route, 59–60
  uploading/read:
    program and comments, 148–149
    project, 63–65
  USB adaptor, 60
  verification, 62–63
Set-reset programming, 175–179
  description, 175
  program/project – PNEU4, 175–179
    ladder diagram, 177
    principle of operation, 177–178
    sequence function chart, 176
    simulation and monitoring, 178–179
Shift registers, 279–281
  applications, 279
  basic operation, 279
  instructions:
    ROR – M8022, 280
    SFTLP – BREW1, 285
    SFTRP, 280
    ZRST, 280
  programs/projects:
    ROTARY1 *see* Index table application
    SHIFT1, 280–281
      ladder diagram, 280
      monitoring, 281
      operating procedure, 281
      principle of operation, 280
Simulation and monitoring:
  ladder logic tester, 214
  PNEU1, 98
  PNEU4, 178
  STEP_CNTR1, 329–330
Simulation Software – ladder logic tester, 214
Single cycle – PNEU1, 91
Software:
  Gx-Developer, xv, 18
  Gx-Simulator, xv, 214
  MEDOC:
    converting *see* Converting MEDOC to Gx – Developer
    preface, xv
Source-sink *see* Inputs
Special M coils, 171–174
  device batch monitoring, 171–172
  list of, 171
  option set up, 173
Stack:
  first in first out FIFO, 333
  instructions:

SFRD – shift read, 335
SFWR – shift write, 334
pointer and registers, 337–338
program/project *see* Automatic queuing system
Start-up procedure, 18–23
  display settings – zoom, 19–21
  drive path, 19
  installation Gx – Developer, 18
  key press numbers – MEDOC format, 22
  main menu, 18
  minimum display of icons, 18–19
  open new project, 19–21
  project data list, 22–23
  project name, 19
Start/open new project *see* Start-up procedure
Statements *see* Documentation
Step counter programming, 324–331
  basic principle of operation, 324
  counter assist, 324–325
  program/project – STEP-CNTR1, 325–331
    instruction – DECO, 326
    ladder diagram, 325–326
    pneumatic panel operation, 331
    principle of operation, 326–329
    simulation and monitoring procedure, 329–330
  sequence control M coils, 324
Step/line numbers, 24, 32
Stepper motor – assignment, 381
Sub-routines, 310–314
  description, 310
  instructions:
    CALL, 311
    FEND, 311
    SRET, 311
  program flow:
    flow diagram, 311
    principle of operation, 311
  program/project – SUB1, 312–314
    labels, 311, 313
    ladder diagram, 312
    monitoring, 313–314
    principle of operation, 311
    sub-routine, 313
Switch bounce:
  A/B phase counter – HSC4, 261
  de-bounce capacitor, 254, 255
  interrupt – INT1, 321, 322

Timers:
  address range, 15
  programs/projects:
    FLASH1 *see* Producing a ladder diagram
    MIXER1, 210–213
      description, 210
      indirect counter setting, 211
      ladder diagram, 211
      monitoring, 212–213
      principle of operation, 211–212
  used list, 39
Trace, 180–194
  block diagram, 180
  description, 180
  principle of operation, 180–181
  program/project – PNEU4, 181–194
    ladder diagram, 181
    obtaining trace waveforms, 190
    reading set up data, 186–187
    saving set up data, 185–186
    set up procedure, 182–185
      conditions, 183–185
      data devices, 182–183
    start trace operation, 187–188
    trace results, 190–194
      calculation time delay T0, 194
      measurement time delay T0, 193–194
      waveform display, 192
    transfer trace data to PLC, 185
    trigger signal – X0, 189–190
Traffic light controller:
  assignment, 378–380
    program/project – TRAF1, 71–74
      description, 71
      ladder diagrams, 72–73
      minimisation modifications, 73
      principle of operation, 74
      truth table, 72
Transfer *see* Serial transfer

Up/down counting:
  COUNT5 *see* Bi-directional counters
  HSC1 *see* High speed counters

Uploading *see* Serial transfer
Used I/O:
  list of used devices, 38–39

V index register *see* Index registers
VDU:
  change of colour display, 134–135
Verification, 62–63
Voltage supplies, 2, 4, 6–7

Waveforms:
  1 second clock signal – M8013, 89
  A/B phase – HSC4, 263
  A/B phase encoder, 249
  decade divider – HSC2, 256
  falling pulse instruction – PLF, 342
  furnace controller – FURN1, 75
  interrupt – INT1, 317
  master control – MC1, 90
  motor controller – HSC3, 258
  queuing system – QUEUE1, 337
  rising pulse instruction – PLS, 83
  RPM meter – REV1, 206, 208
  stepper motor assignment, 381
  timing chart – LLT1, 225
  trace – PNEU4, 190–193
Weighting, binary to decimal/denary, 196
Wiring diagrams:
  emergency stop, 113
  high speed counter – HSC4, 261
  PNEU1, 100
  simulation unit, 97

X inputs *see* Input

Y outputs *see* Output

Z index register *see* Index registers
Zoom *see* Start-up procedure